The Economics of the Wildlife Trade

This book presents a thorough overview and assessment of the economics of the wildlife trade for scholars and conservationists, requiring only a basic understanding of economic principles.

Wildlife conservation and the sustainable extraction of living organisms from the wild for human use are fundamentally tied to issues of resource allocation and economics. Yet, there has been a conspicuous dearth of scholarly contribution that examines the wildlife trade using the theories and analytical tools provided by economists. This book fills this gap by presenting a theoretical framework built on a foundation of microeconomic theory that is used to analyze the trade in wildlife goods, the potential effects of various conservation policies, and the economic forces that govern the workings of the markets for wildlife goods. The focus of the book is on explicating the possible impacts of supply-side policies that have been implemented or proposed to achieve a sustainable level of wildlife extraction. The policies that will be considered include the following: stricter anti-poaching policies; trade bans; and legalizing the sale of alternative sources of supplies such as farmed goods, synthetic substitutes, and accumulated stockpiles. The book discusses how the implementation of these policies could affect the incentives of both the suppliers and buyers of wildlife goods to participate in these markets. It also considers how market structure, market concentration, and asymmetric information can affect the impact of these policies.

This book is intended for scholars, practitioners, and policymakers in the field of wildlife conservation who are familiar with basic economics and interested in a rigorous exposition of the economic principles relevant to the study of the wildlife trade.

Frederick Chen is a Professor in the Department of Economics at Wake Forest University, USA. He teaches microeconomics, mathematical economics, and game theory. His current research interests lie mainly in the intersection of economics and population biology.

Michael 't Sas-Rolfes is an interdisciplinary conservation scientist and economist with four decades of experience in the wildlife sector. He is a Research Fellow at the Interdisciplinary Centre for Conservation Science and the Oxford Martin Programme on Wildlife Trade at the University of Oxford, UK, and at the African Wildlife Economy Institute at Stellenbosch University, South Africa. He serves on the IUCN Species Survival Commission's African Rhino Specialist Group and its Sustainable Use and Livelihoods Group.

Routledge Studies in Conservation and the Environment

This series includes a wide range of inter-disciplinary approaches to conservation and the environment, integrating perspectives from both social and natural sciences. Topics include, but are not limited to, development, environmental policy and politics, ecosystem change, natural resources (including land, water, oceans and forests), security, wildlife, protected areas, tourism, human-wildlife conflict, agriculture, economics, law and climate change.

The Maya Forest Waterlands
Shared Conservation, Entangled Politics, and Fluid Borders
Hanna Laako and Edith Kauffer

Saving Biodiversity
Threats, Strategies, and Big Ideas
Matt W. Hayward

Rewilding and Ecological Justice
The Ethics and Politics of Wildlife Regeneration
Cristian Moyano-Fernández

Becoming Nature Positive
Transitioning to a Safe and Just Future
Marco Lambertini, Joseph W. Bull, Leroy Little Bear, Harvey Locke, Eva Zabey, Dorothy Maseke, and Carlos Manuel Rodríguez

The Economics of the Wildlife Trade
Theory and Implications for Conservation
Frederick Chen and Michael 't Sas-Rolfes

For more information about this series, please visit: www.routledge.com/Routledge-Studies-in-Conservation-and-the-Environment/book-series/RSICE

The Economics of the Wildlife Trade

Theory and Implications for Conservation

Frederick Chen and Michael 't Sas-Rolfes

Routledge
Taylor & Francis Group
LONDON AND NEW YORK

earthscan
from Routledge

Designed cover image: Michael 't Sas-Rolfes

First published 2026
by Routledge
4 Park Square, Milton Park, Abingdon, Oxon OX14 4RN

and by Routledge
605 Third Avenue, New York, NY 10158

Routledge is an imprint of the Taylor & Francis Group, an informa business

British Library Cataloguing-in-Publication Data
A catalogue record for this book is available from the British Library

ISBN: 978-1-032-26057-0 (hbk)
ISBN: 978-1-032-26053-2 (pbk)
ISBN: 978-1-003-28631-8 (ebk)

DOI: 10.4324/9781003286318

Typeset in Optima
by Newgen Publishing UK

Access the Support Material: www.routledge.com/9781032260532

Contents

Preface

During the first quarter of the twenty-first century, wildlife trade resurged as a prominent concern for global public policy. The final quarter of the twentieth century had seen increasing trade regulation under the banner of the Convention on International Trade in Endangered Species of Wild Fauna and Flora (CITES), which was established to support positive species conservation outcomes in the face of growing commercial pressures on many wild species. However, even after five decades, the effectiveness of CITES continues to be questioned, given mixed results and a notable surge in the illegal killing of prominent charismatic species such as elephants, rhinos, and pangolins in the early 2000s. Wildlife trade—and the appropriate ways to govern it—remains a hotly debated and heavily polarized topic at the time of writing this book.

We aim to contribute to the discussion around the topic by providing insights that we both have gained from the discipline of economics. One of us (FC) is an economic theorist who has moved from deep engagement with the world of theory to apply his mind to practical problems such as public health and wildlife trade policy. The other (MtR) started out as a "pracademic," linking both practical ecological and commercial experience with insights gained from academia to engage in a passionate interest in wildlife conservation. We share a desire for effective environmental governance and a belief that economic thinking can provide helpful guidelines for achieving this.

In our experience, many conservation scientists, regulators, and people who care about wildlife are suspicious of trade and economics, perceiving associations with greed (pursuit of money) and excessive consumption. We believe that this perception arises partly from the limited exposure to economic thinking provided by training in disciplines such as conservation biology and environmental law, scholars of which have been the most influential in determining wildlife trade policy. With this book, we hope to provide a fresh perspective that will enable us all to work together toward achieving better outcomes for both people and wildlife.

We have attempted to write the book for a broad audience with varying knowledge of scholarly economics. Readers will ideally have a basic understanding of economic concepts and principles, but even those without such knowledge should benefit from reading parts of this book. In essence, the book consists of three parts. Chapters 1–3 constitute the first part and provide essential background for understanding the topic of wildlife trade and conservation, introducing key terms and concepts. Chapters 4–7 constitute the second part, which is the heart of the book, presenting the rigorous economic theory that forms our foundational approach to the topic. These chapters require greater concentration from those unfamiliar with basic economics, especially as the analyses are expanded. Chapters 8–10 constitute the third part of the book, in which we step back from the theory somewhat and ask the question: "So what"? In this part, we pull together the insights from the first and second parts to consider how they might be applied toward finding more effective ways to tackle wildlife trade for the benefit of conservation.

Although most readers ought to benefit from the entire book, some may wish to focus on specific parts. While all readers should benefit from the first part, scholars interested in learning the economic theory will benefit the most from a deep dive into the second part. Conservation practitioners and policymakers may find they gain more from concentrating on the third part, which highlights the practical and policy implications of our economic approach.

More advanced economics scholars may find the economic treatments somewhat elementary—we have mostly avoided the complexity of dynamic analysis and its associated calculus. However, the appendices of Chapters 5–7 provide some additional technical detail for those wishing to explore further. There remains much scope for further detailed investigation of specific aspects, and we hope that the issues we have raised will stimulate avenues for further theoretical and empirical research. Above all, we hope that our contribution will broadly advance the knowledge and understanding of this important topic.

1 Introduction

Trade in wildlife products has played a significant role in human society throughout history and continues to feature prominently in the modern global economy. Some aspects of this trade greatly benefit humanity, but poorly governed trade can also cause environmental and social harm. Humans should understand and engage appropriately with wildlife trade to minimize such harm. This book aims to deepen our collective understanding of the relationship between wildlife trade, policies toward it, and various outcomes, especially consequences for conservation. This chapter commences with an overview of the wildlife trade and its potential positive and negative effects. We explain why we view problems associated with this trade as essentially economic and how economic analysis can deepen our understanding of them. After providing some basic information about economics and key insights from economic history, we discuss some areas of confusion in wildlife trade policy literature (academic and other) that can be clarified through the appropriate use of economic analytical tools. The book proceeds to build a framework with such tools.

Wildlife Trade: Definitions and Issues

Before discussing wildlife trade, we need to define it, which is not as straightforward as one might think because neither of the terms—"wildlife" and "trade"—has a universally applied meaning. Here, we briefly explain why and then elaborate further in the final section of this chapter on how we intend to treat the scope of this topic. Having defined wildlife trade, we can identify some significant issues of interest.

Definitions

As a simple definition, "wildlife" typically refers to animals and plants that have not been domesticated by humans and live in "natural environments" ("in the wild"). Everyone would likely agree that elephant herds roaming in Africa's large parks or penguins in Antarctica are wildlife and that poodles kept

DOI: 10.4324/9781003286318-1

as house pets or Jersey cows on a dairy farm are not. However, some people only think of animals as wildlife and not plants (or fungi, which are now recognized as a separate third category). Furthermore, even among animals, there are many grey areas: for example, (1) formerly domesticated animals that have become feral (e.g., horses and cats), (2) species that are typically thought of as wild but are now also farmed (e.g., alligators and ostriches), and (3) various antelope species that occur on wildlife ranches in Southern Africa, where enclosures range in size from paddocks to fenced private reserves that are larger than many state parks. Leaving the grey areas aside for now, we regard "wildlife" as comprising all the (millions of) animal, plant, and fungus species that continue to exist at least partly outside of human-dominated contexts (i.e., all but a relatively short list of less than 50 animals, 50 fungi, and several hundred plants).

"Trade" is seemingly easier to define as the voluntary exchange of goods and services, usually (but not always) for money.[1] In this context, a "good" is anything that is valued by humans and can be owned (i.e., viewed as "property"), and a "service" is any valued action or function that humans or human-owned entities can provide for the benefit of others. Studying trade is a cornerstone of the discipline of economics. A trading relationship is distinct from one of altruism (where one party gives something with no expectation of return) and one of involuntary transfer (in which one party gains a good or service at another's expense, through coercion, theft, or fraud). Although other animal species may engage in some basic forms of reciprocal exchange, the human species has evolved to trade over long distances across the world, with goods potentially changing hands numerous times along the way. The concept of "property" as a universal custom also appears to be uniquely human (Wilson, 2020). Paleoanthropological evidence indicates that humans have been trading goods ever since we emerged as a species[2] (Wolff, 2023) and that early traded goods included items fashioned from wildlife products such as seashells and the bones of hunted animals. The entities through which trade occurs are called "markets."

"Wildlife trade" obviously refers to the trade in wildlife, but this can take many forms, some more obvious than others. Confusing the issue somewhat is the question of wildlife "ownership." Although humans have agreed to recognize wildlife as tradable property in some instances, in others, they have not or have otherwise specified limited conditions under which ownership can be established. For example, Roman-Dutch law treats wildlife as *res nullius* (owned by no one) until it is killed, captured, or domesticated; beneficial ownership is thus only recognized after one of these has happened. Obvious forms of wildlife trade involve exchanging individual animals or plants taken from the wild (dead or alive) or parts of them, such as animal skins and plant seeds. Less obvious forms of trade may occur when ownership of land with wildlife on it is transferred, when people can control access to (and therefore sell) wildlife viewing opportunities, or when people take images of wildlife (photographs or films) and use those to earn money.

Human concern about wildlife trade tends to focus on the obvious forms that involve the visible extraction and handling of wildlife specimens, and these are also the focus of the Convention on International Trade of Endangered Species of Wild Fauna and Flora (CITES), a treaty between national governments (the "Parties") that came into force in 1975. Through CITES, the Parties agree to create and enforce laws and regulations to restrict or otherwise influence trade in specimens of selected wild species. The existence of CITES—and its regulatory logic—has also shaped the way many people think about wildlife trade. CITES' own definition of "trade" is somewhat different from that typically used by economists because it relates to humans transporting wildlife specimens across international jurisdictional boundaries, even without a transactional change of ownership.[3] The Convention text also refers to trade for "primarily commercial purposes" and distinguishes this from "non-commercial" trade-related activity, creating confusion and leading to varying and potentially conflicting interpretations among the Parties.[4]

Ignoring the confusion this creates for now, the CITES approach also highlights two other aspects of trade that receive regular attention: legality and sustainability. Legality of wildlife trade, i.e., whether it is legal or illegal, is simply defined by whether it complies with the laws and regulations established by governments. Sustainability is less well defined. A sustainable wildlife harvest is one that does not cause the decline of the harvested species in the wild over time. References to "unsustainable wildlife trade" in the literature imply that this type of trade is linked to unsustainable rates of extraction through harvest that result in species depletion. However, the causal connections between trade activity and extraction rates are not straightforward, as discussed below and in Chapter 2. Finally, it is important to note that legality and sustainability are not necessarily correlated. Legal trade can sometimes be linked to unsustainable harvesting, and not all illegal trade threatens species ('t Sas-Rolfes et al., 2019).

Issues

Economists mostly view trade as positive for human development and well-being to the extent that it does not cause harmful side effects, and this can include wildlife trade. Wildlife products provide many positive benefits to humanity. Before humans domesticated certain species, our ancestors depended on wildlife for food and used wildlife products for fuel and building materials, as well as for making tools, clothing, ornaments, musical instruments, and medicines. As humans domesticated species and congregated in urban areas, their dependence on wildlife declined. However, as human populations also expanded and colonized new areas, they sometimes harvested local populations of wild species to extinction. Scientists can only speculate over the extent to which human hunters caused the extinction of large animals (megafauna) in the Pleistocene Era (Barnosky et al., 2004), but there are documented examples of this in more recent history—for example, the case of the bluebuck, a species eliminated by colonial settlers in the Cape region of South Africa by the early nineteenth century.

While many historical instances of wildlife being harvested to extinction may have been related to direct consumption or use by those doing the harvesting, the mid-nineteenth century provides a classic example of overexploitation linked to international trade: the American bison. With their numbers and range already affected by encroaching human settlement and hunting, large-scale slaughter of bison herds took place in the 1870s after tanners in Europe developed a method for tanning bison hides into useful leather, ultimately to make belts and shoe soles (Taylor, 2011). Within a decade, the continental bison population was reduced from an estimated 10–15 million animals to near extinction. The near loss of this species helped prompt the rise of the nature conservation movement in the late nineteenth and early twentieth century, which was accompanied by the first attempts by governments to regulate cross-border wildlife trade, being the Lacey Act in the USA, and a London Convention for the Preservation of Animals, Birds and Fish in Africa, both established in the year 1900.

Early concerns about wildlife trade were focused on overexploitation, with wild species viewed as valuable renewable natural resources that should be maintained by preventing unsustainable harvesting rates. Hence, the main objective of wildlife trade regulation was to discourage extractive harvesting to prevent species from going extinct. However, there are other reasons to be concerned about wildlife trade than its associated extraction rates, and these are increasingly highlighted in recent literature (e.g., Hughes et al., 2023). Trade can act as a vector for invasive species to be introduced to ecosystems where they previously did not occur and where, without natural predators, they can cause other local species to go extinct and create other environmental problems (Blackburn et al., 2014). Selective harvesting for trade can also negatively impact species and ecosystems through various other indirect mechanisms, such as adversely affecting desirable traits, population demographics, trophic cascades, seed dispersal, predation, and carbon stocks. In addition to alien invasive species, wildlife trade can enable the spread of disease-causing pathogens that can infect humans, livestock, and native wildlife populations, causing substantial ecological and economic damage (Karesh et al., 2005), a concern that was highlighted by the COVID-19 pandemic (Roe et al., 2020).

Many people also care about animal welfare and how trade may negatively impact it. Examples of concern include how humanely wild animals are harvested (whether live or killed), how traded live animals are transported, the treatment of captured live animals destined for slaughter, and the treatment of animals kept in zoos or as pets (Baker et al., 2013). Animal welfare and related moral sentiments feature increasingly in trade policy debates, underpinning campaigns against wildlife commodification ('t Sas-Rolfes & Gooden, 2024). Animal welfare is important not only because it arouses emotions in people but also because poor welfare conditions increase susceptibility to infectious diseases and thus enable disease to spread, an issue of growing concern to international policymakers (Goetschel, 2024).

When wildlife use and trade are banned or regulated, some people may ignore or willfully break the rules. Such infractions include the illegal taking of animals, plants, and fungi from the wild (i.e., "poaching") and the illegal trading and transporting of wildlife products, often referred to as wildlife "trafficking." Together, poaching and trafficking[5] are often referred to as "wildlife crime." The existence of wildlife crime creates further problems to add to those already mentioned, which may also become harder to address. Since the inception of CITES, both the numbers of regulated species and the extent of observed wildlife crime have increased significantly worldwide, not only amplifying the previously mentioned environmental and animal welfare problems but also causing substantial harm to people. For example, the illegal trade in more lucrative products, such as rosewood timber, rhino horn, elephant ivory, pangolin scales, and abalone, has attracted and fueled dramatic increases in transnational organized criminal activity and associated violence. Many human lives have been lost or otherwise severely affected by violent interactions between poachers and rangers in the field.

During the early twenty-first century, the growing scale and intensity of this criminality led to the United Nations flagging it as more than an environmental concern to one negatively impacting economies, livelihoods, good governance, and the rule of law and, hence, becoming an issue of national and international security (Nellemann et al., 2014). The United States government embraced and promoted this reframing of wildlife trafficking as a security issue with geopolitical dimensions, prompting criticism and expressions of concern over the consequent "militarization" of conservation by social scientists (Duffy, 2016; Massé and Margulies, 2020; Duffy et al., 2019). Local rural communities living with wildlife have increasingly been caught in the crossfire of the escalating war against wildlife crime, resulting in numerous human rights abuses in the name of conservation (Duffy, 2022), and prompting some conservation social scientists to call for an alternative "socioecological harm reduction" approach (Hübschle & Margulies, 2024). Some of the issues relating to wildlife trade arise from how policymakers and governments approach it.

Wildlife Trade and Economics

The discipline of economics has evolved over time and therefore been defined in various ways; economists continue to disagree over its nature and scope (Wilson, 2025). Classical approaches to economics view it as concerned with how and why humans engage in trade to create wealth, as exemplified in Adam Smith's seminal work "An Inquiry into the Nature and Causes of the Wealth of Nations" in 1776, and consider it intertwined with ethics and politics. More recently developed "neoclassical" approaches tend to be more technical, i.e., studying rational choice in the face of scarcity and viewing that as something that can be analyzed in isolation and mathematically modeled. Neither of these approaches is intrinsically right or wrong, and both provide valuable insights. Importantly, all approaches to economics consider how markets work through the interactions of the supply of and demand for goods and services.

In most contemporary descriptions, economics considers how humans allocate scarce resources among competing needs and wants. Some definitions of economics specify things like the production, distribution, and consumption of goods and services, and many people associate economics with the study of money and finance. However, economics is broader than that—it is a social science that examines human behavior in certain contexts. We could argue that a person deciding how to allocate their limited and valued time between two competing options is making an essentially economic decision. In broad terms, both options will have expected pros (benefits) and cons (costs) that may be compared. Typically, the person would choose the option they expect to yield the greater net benefits (benefits less costs) to them, also noting that this means forgoing the net benefits of the other option, which economists call the "opportunity cost." Economics provides a way of thinking about trade-offs, which are a ubiquitous feature of everyday life. The logic of everyday life decisions—weighing up benefits and costs and responding to certain incentives—applies as much to the decisions humans make about wildlife trade and conservation as they do to other topics that are more conventionally considered to be within the economic domain.

Wildlife Economics

Conserving wildlife is not a straightforward pursuit. In Chapter 2, we discuss how complicated it can get, but for now, let us focus on a core objective: to prevent any individual species from becoming extinct. The problem we face is that species compete with other species, including humans, for things that we all need, notably space, shelter, energy intake (including sunlight and food), and water. Humans also compete among themselves for the same things. We can think of conservation as aiming to find ways for humans and all other life forms to all receive a reasonable share of these things, to stay alive and, hopefully, to also experience a good-quality life.

However, as we have seen from the remarkable human population growth that has occurred in recent history, this has greatly increased the numbers of relatively few (now mostly domesticated) species that are useful to us and resulted in reduced numbers of most others. An analysis of the earth's mammalian biomass in 2015 estimated that 34% was made up of humans, 62% was made up of domestic livestock (cattle, pigs, sheep, goats, horses, etc.), and only 4% was made up of all other wild species, including both land and marine mammals (Ritchie & Roser, 2024). Wild species have lost space and other resources to competition with humans and their useful livestock. A person living in comfort in a modern city may find this sad and feel a strong urge to strictly protect the remaining wild species. However, for people living in rural areas and depending directly on the land for their livelihood, their experience of wildlife—and their attitude toward it—can be very different.

Understanding the lived reality of people struggling to survive on the rural frontier is key to understanding the economics of wildlife conservation. Rural subsistence farmers depend on their family members, livestock, and crops for their livelihoods. When such people live alongside or among wild species, they tend to view such species through the lens of either supporting or threatening their existence. When those wild species include large dangerous animals such as elephants and big cats (which may destroy their crops or attack and kill their family members and livestock), they face the prospect of losing their lives or livelihoods in a single overnight raid or predation event. Even less dangerous wildlife may be a serious nuisance, such as other grazing mammals competing for forage with their livestock, other predators taking their smaller animals (e.g., lambs and chickens) and eggs, and birds and insects eating their crop plants. In other words, tolerating the presence of wildlife often incurs both direct and opportunity costs.

Over time, unless rural people receive sufficient net benefits from the continued presence of wildlife to offset the costs, they will tend to eliminate the costly species and replace them with species that (1) they can legally own and (2) provide them with greater net benefits—typically crops and livestock. This process has happened throughout human history and explains why developed and densely populated parts of the world (e.g., Western Europe) have relatively little wildlife, especially large mammals. The African country of Kenya provides a recent illustrative case study. Long renowned for its spectacular wildlife, including large charismatic animals such as elephants, lions, and rhinos, Kenya banned all hunting of large wild mammals in 1977. However, from 1977 to 2016, the country experienced extreme losses of these animals (on average, 68%) in their rangelands (Ogutu et al., 2016). The unfolding of this decline was documented and explained by Mike Norton-Griffiths (1996, 2007), an ecologist who subsequently studied economics and linked the losses to Kenya's human population growth and development trajectory. Following the decline, Kenya has begun to reevaluate its official conservation policies and explore new ways for its citizens to benefit from the existence of wildlife.

To the extent that people perceive wildlife to be valuable to them—and it provides them with meaningful direct net benefits—they are far more likely to expend the effort to conserve it, from local to national government levels. However, the question of how this value is provided, experienced, and shared within society is critical. In his popular book *Sapiens*, the historian Yuval Noah Harari argues that humans coordinate their actions through "imagined orders" such as religions and governments but highlights the role of money as "the most universal and most efficient system of mutual trust ever devised" (Harari, 2014: 180). Wildlife conservation cannot succeed fully without recognizing and engaging with the extent to which humans coordinate their collective efforts through various forms of trade, typically involving some form of money. Economics provides us with the tools to analyze this and provide guidance for more effective conservation policies.

Lessons from Economic History

Gaining an in-depth understanding of economics takes time and requires much study. However, by briefly examining contributions from economic history, we can identify a few key insights from leading thinkers that we can apply to the wildlife trade. The first documented musings about economic issues date back to the Greek philosophers Plato and Aristotle, who already had differing views but were both interested in questions concerning property, exchange, and specialization (being foundations of economic analysis) in relation to community, self-sufficiency, and justice (Ekelund & Hébert, 1983). These issues were examined by other Western philosophers from the Middle Ages through to the eighteenth century, at which time Adam Smith's writings raised the public profile of economic thinking.

Smith argued that humans have an inherent propensity to "truck, barter, and exchange one thing for another" and that the public interest was served by people specializing in certain activities and trading in competitive markets, which acted as an "invisible hand" to guide people acting in their own interest toward outcomes from which broader society would benefit. The "invisible hand" was Smith's metaphor for the force that drove the competitive interactions of sellers (supply) and buyers (demand) toward finding a socially optimal "equilibrium" point that would determine the market price of a good or service. Smith also claimed that monopoly power in markets would distort market prices and typically lead to less socially desirable outcomes. Smith thus argued a case for unregulated (and therefore competitive) "free" trade, receiving further support in the early nineteenth century from David Ricardo, a successful trader who presented the theory of comparative advantage,[6] a further compelling argument that humanity benefits from specialization and free trade at an international scale.

In 1890, Alfred Marshall contributed significantly to coherent economic theory with his book *Principles of Economics*, which further developed and synthesized many foundational concepts, including elasticity, increasing and diminishing returns, marginalism, consumer surplus, externalities, and differences between short- and long-term phenomena.[7] By illustrating supply and demand graphically as curves relating prices to product quantities, Marshall demonstrated (1) the significance of the equilibrium ("market clearing") point and (2) the relevance of elasticities (reflected by the slopes of the curves) in determining the sensitivities of the price–quantity relationship. Marshall also expanded his remit of economics beyond a study of "wealth" to include a broader concept of human "welfare," which prompted Lionel Robbins (in the early 1930s) to propose a revised (and now widely accepted) definition of economics as a science concerned with human behavior under the influence of scarcity.

During the twentieth century, diverse strands of economic thought continued to evolve, somewhat shaped by the experience of various significant events such as the two World Wars, The Great Depression, and the fall of the Iron Curtain. Several noteworthy figures, all of whom won Nobel prizes for their contributions, provided insights that we consider relevant to our topic. An article titled "The Use of Knowledge in Society" by Friedrich Hayek (1945) provided a convincing critique of central economic planning, explaining how the decentralized market pricing system works as a far more effective way to coordinate dispersed information about scarcity than any government could. The contributions of Ronald Coase highlighted the foundational role of both property rights and transaction costs in shaping the pathways via which social, economic, and environmental problems ("market failures") could be resolved.[8] Later work by others would affirm that poorly defined or weak property rights are at the root of many environmental problems.

Gary Becker applied economic thinking to broader social issues, providing noteworthy insights into the rational incentives to commit crime and the effectiveness of varying deterrent measures (Becker, 1968). In a similar vein, James Buchanan considered that politicians and public officials may act as rational agents with private interests (facing economic incentives that often misalign with public interests) and accordingly co-developed and investigated a theory of "public choice." More generally, by studying historical data, Douglass North (1990) and others demonstrated that institutional arrangements critically affect economic performance. Institutions, which we loosely define here as the rules by which society organizes itself,[9] create incentives that shape human behavior, which determines outcomes. In the early twenty-first century, the research of Daron Acemoglu, James Robinson, and Simon Johnson further affirmed the significance of varying institutions to economic prosperity, earning them the 2024 Nobel Prize in Economics.

In 1968, Garret Hardin, an ecologist, drew on the 1832 work of British economist William Forster Lloyd to write the famous article "The Tragedy of the Commons" (Hardin, 1968), which argued that if people enjoy unrestricted open access to a shared resource that is finite and valuable, they will tend to overuse it and might destroy it completely. Hardin's article was subsequently criticized for creating the impression that all common property regimes suffer from this issue, whereas there are many examples of successful communal resource management through collective action. A large body of work by Elinor Ostrom[10] and others has since developed sophisticated theories of property rights and institutional design for effective common property and natural resource governance, informed by applying systems thinking to explore the interactions of people and ecosystems. Ostrom and her colleagues developed a pragmatic research approach to ensure that it was not purely theory-driven but rather informed by iterative referral to real-world outcomes. This integrative work provides the context for understanding how and why economics can provide critical insights for designing effective wildlife trade policy and governance systems that address the issues of concern that we raised in the previous section.

Misconceptions about Wildlife Trade

Returning to the issues of concern relating to wildlife trade, and especially the intention to prevent the extinction of wild species populations, it is worth reflecting on why many observers lament the current situation.[11] Tracing the evolution of CITES and associated attempts to tackle wildlife trade problems, we can observe many instances of naïve understanding of the economic forces that drive trade (legal or illegal) and how to deal with them. To illustrate this, consider the example of rhinoceros (hereafter "rhino") conservation.[12] Apart from a relatively brief lull period, all five rhino species have faced a tough time since the early 1970s, being widely subjected to intense poaching pressure ('t Sas-Rolfes et al., 2025). A decades-long campaign to remove this threat through measures aimed at regulating or preventing trade can claim only limited success, and the threats to rhinos persist. Why is this?

Humans have used (and traded) rhino horn for many centuries, even millennia, for both ornamental and medicinal purposes. Rhino species and populations have also declined over this time (Liu et al., 2021) for various reasons, including natural climate change and anthropogenic (human-caused) factors such as habitat conversion and hunting for meat and other products (not just horn). In the early twentieth century, previously depleted rhino populations started to recover within protected areas, but the early 1970s saw an economically relevant event not unlike what happened with the American bison a century earlier. In this case, the catalyst was the global oil crisis, which indirectly brought unprecedented prosperity to Yemeni workers in Saudi Arabian oilfields, fueling their culture-based demand for traditional ornamental dagger handles made of rhino horn. This spike in demand added to an existing East Asian market for powdered rhino horn as a traditional medicinal ingredient, causing horn prices to rise and thus stimulating a wave of rhino poaching across Africa.

This event coincided with the establishment of CITES, and rhino species were immediately listed as threatened and therefore subject to international trade restrictions. At the time, prominent conservationists claimed that "trade" threatened rhinos and thus simply needed to end. To achieve the termination of trade, several measures were proposed to "close" markets as a supplement to existing anti-poaching measures. The literature around that time reveals that no distinction was made between reducing supply and reducing demand, nor in what sequence these goals should be pursued. Any attempts at stifling trading activity were considered desirable—a mindset that persists today—and cutting off any legal supply was the first step toward doing this. The problem with this approach (as studying economics has taught us) is that reducing a product's legal supply without adequately considering and addressing consumer demand[13] typically raises its market price. For a wildlife product, if there is no commensurate increase in the cost of poaching and illegal trade, the increased price provides a greater economic incentive for suppliers to provide the product illegally, and this acts as a more potent driver of poaching.[14] Following the initial attempts to restrict legal trading activity, market prices for rhino horn did indeed increase dramatically (notably relative to the costs of illegal supply), and it was only after influencing rhino horn consumer countries (in the early 1990s) that price increases and poaching pressure eventually subsided, albeit temporarily (Vigne et al., 2025).

Implicit in a more restrictive regulatory approach toward trading activity is a set of assumptions, which has proven to be weak in the case of the rhino horn trade (and that of many other wildlife products). A critical assumption is that all the Parties to CITES (and the regulatory agencies within them) will cooperate and implement this approach to the extent necessary. However, there are many (somewhat interrelated) reasons why they may not. As an overriding principle, any laws that are passed should have the overwhelming support of society if they are to be effective. In other words, legality should be mostly aligned with social legitimacy. Social science research on the resurgent twenty-first-century rhino horn trade showed that this is not the case, with the law being contested by many different people all along the supply chain, from poachers and wildlife custodians through illegal traders to consumers, including government officials at all levels (Hübschle, 2016).

Given that laws may be contested or ignored, effective CITES implementation must at least rest on the assumption that governments will do their part to pass relevant domestic laws, establish the agencies to implement and enforce them (including relevant conservation bodies, police, prosecutors, courts, and correctional service providers), provide those agencies with sufficient financial and other resources to run effectively, and ensure that all enforcement agents are incentivized to do their job. In many instances, one or more of these elements is missing, so the assumptions are not met (Challender et al., 2025). This is often blamed on a "lack of political will." However, closer economic analysis drawing on principles of public choice theory would likely reveal that the governments of many less affluent countries simply cannot afford to do all these things effectively, especially when facing many other pressing political and economic demands from their citizens. In such situations, opportunities for corruption of government officials also abound (Smith & Walpole, 2005).

The relative inability or unwillingness of many countries to fully enforce restrictive wildlife trade regulations leaves conservationists hoping that they can persuade people to change their behavior through appeals to morality and ethics. However, morality varies with culture and other circumstances, and people may be motivated to continue using wildlife products for wide-ranging reasons (Thomas-Walters et al., 2020). We can envisage that a person considering whether to harvest or consume a wild organism might simply assess whether the benefits outweigh the costs.[15] They may not consider the activity to be immoral to the extent that it constitutes an intolerable perceived ethical cost (if at all). Furthermore, if they know the activity is illegal, they can include the expected cost of being caught and punished as part of their assessment. If the penalty is not sufficiently high or (more importantly) the probability of incurring it is low, a rational economic calculation might suggest that committing a crime is worth it (especially for people with limited other options). This has been clearly demonstrated by research into rhino and elephant poaching in Zambia (Milner-Gulland & Leader-Williams, 1992).

The long-term failure to reduce rhino poaching to comfortable levels by outlawing trade activity has prompted alternative proposals to increase the supply of horn or provide horn substitutes through various legal means (Di Minin et al., 2022). Such approaches have been adopted successfully for many other CITES-listed species, notably including crocodilians (Hutton & Webb, 2002) and the vicuña, a South American camelid that yields prized wool (Kasterine & Lichtenstein, 2018). However, in the case of rhinos, there has been much skepticism and many counterarguments. Caution is justified, but some of the counterarguments are economically incoherent. For example, some skeptics claim that the demand is simply "too large" for the potential supply. This argument ignores how supply and demand curves interact and how the quantities traded are modulated through prices. Another argument suggests that any increase in supply from an existing legal and sustainable source of a wild-harvested product (or even the mere act of trading it) will inevitably "stimulate" (greater) increased demand for illegally harvested products as a response. This argument implies that there is an asymmetric causal relationship between legal quantities supplied (or traded) and the desire to purchase illegal products, again ignoring the role of prices and making dubious assumptions about consumer demand—how it originates and grows, and the willingness of consumers to opt for known legal and sustainable sources. These arguments are not supported by a basic economic understanding of how markets work.

Moving on to more plausible and coherent concerns, skeptics argue that the act of legalizing a previously illegal trade might bring new consumers into the market (thereby expanding it) and that adding a legal supply channel opens the way for product "laundering" (i.e., providing an easier and cheaper way to bring illegal products into the market by sneaking them into the legal market), which complicates law enforcement and could incentivize further illegal activity.[16] Although these are legitimate concerns for conservation under certain conditions, under others, they are not, as we shall set out to explain in this book. As an extension of these concerns, some trade opponents argue that accumulated stockpiles of wildlife products, such as rhino horn and elephant ivory, should be destroyed to prevent them from potentially entering the market, a proposition that conflates trading activity levels with rates of extraction from the wild, thereby reinforcing the misconception that any "trade" whatsoever necessarily leads to a bad conservation outcome. Finally, trade opponents argue that alternative products supplied either by wildlife farming or the production of synthetic substitutes will not displace wild-extracted products because consumers always prefer the latter. Not only has the argument been debunked by consumer surveys that show widely varying preferences, but even when wild products are preferred, introducing substitutes may have a positive conservation impact under certain circumstances.

Much of the opposition to the legal trade of wildlife products centers on ideas that "trade" is somehow inherently bad for conservation, in some cases evidently arising from moral intuitions and normative sentiments toward wildlife commodification ('t Sas-Rolfes & Gooden, 2024). However, trade itself is neither intrinsically "good" nor intrinsically "bad," but just something that humans do, and the use of money as a medium of exchange does not intrinsically change the moral virtue of the underlying activity (Brennan & Jaworski, 2015). To the extent that wildlife trade is associated with the various problems we outlined above, these are incidental rather than intrinsically caused by trade. Furthermore, criminalizing (banning) trade does not stop trade entirely; instead, it modifies the incentives that people face and how trade may be conducted, which may or may not solve problems related to conservation or animal welfare. Note that economics does not necessarily prescribe legal trade to address such problems. Rather, it helps provide insights into how rational people are most likely to respond to the incentives created by various laws, policies, and market conditions. Economics is neither immoral nor amoral but can help to reflect the outcomes of differing interacting moral values and preferences and their expression through markets.

If species conservation is our primary goal, then the relationship between some measure of "trade" and the rate of extraction from the wild should be our core analytical focus. Unfortunately, as we have already noted, definitions of trade differ, and the methods that many researchers have used to measure it do not necessarily tell us enough about how trading activity links to extraction rates—we find many methodological errors and inconsistencies in the scientific literature (Challender et al., 2022). In the next chapter, we delve deeper into this question by discussing bioeconomic theory and explaining why we consider property rights, price dynamics, and profitability as more relevant determinants of extraction rates. However, we first conclude this chapter by outlining the scope and content of the rest of this book.

Scope of This Book

We have loosely defined wildlife trade as the voluntary exchange of wild animals, plants, and fungi by humans. Although our definition would recognize secondary (indirect) trade that might occur through, for example, exchanges of land with resident wildlife, this book will focus on the visible commercial trade of physical wild-extracted specimens that is the focus of CITES and supporting domestic regulations. It is this trade—and the extraction that takes place to supply it—that raises the concerns about species conservation, animal welfare, disease spread, and other social spillover effects that we discussed earlier. Noting that economics is a broad discipline with diverse schools of thought, our approach presents a basic economic framework, which we build upon to explore specific issues. This basic framework draws on the foundations of much historical and contemporary economic thought and analysis and seems the most appropriate place to start. However, before introducing the framework, we provide further background in the following two chapters.

Chapter 2 serves as a more detailed review of existing literature and knowledge of wildlife conservation and trade from theoretical and practical perspectives. We delve deeper into the complexities of conserving wildlife, further explain the relevance of institutions in influencing conservation outcomes, and discuss the evolution of thinking about wildlife trade regulation. Chapter 3 describes the elements of markets for wildlife goods, starting with an explanation of how they develop and then describing the components of a typical wildlife product trade chain and the various roles of people ("actors") within it. These descriptions, illustrated by real-world examples, enable better visualizations of the market elements discussed in the subsequent chapters.

Chapters 4–7 comprise the core theoretical content of our book. Chapter 4 introduces our basic economic framework, which builds on the concepts of demand and supply and how they interact through the market price mechanism. We consider the role of market intermediaries (traders) and the consequences of particular market structures and policies. In Chapters 5 and 6, we develop the framework models further to examine the effects of introducing substitute goods under two different scenarios: in Chapter 5, we do so under conditions of complete information, and in Chapter 6, we look at unevenly distributed information. In both instances, we consider the implications of adopting specific policies. Chapter 7 adds a new dimension to our analysis by considering how stockpiling storable wildlife goods may influence market behavior. The potential to store some wildlife goods opens the way for strategic and speculative economic behavior with potentially important consequences for conservation, and this also has policy implications.

Drawing on the insights from the theoretical chapters, Chapters 8 and 9 lay out an economic approach to aligning wildlife trade with conservation objectives. Chapter 8 considers how the incentives of actors along the trade chain may be influenced (focusing on those responsible for the initial extraction of wildlife from its natural habitat) and the implications for effective responses to deter unsustainable harvesting. Chapter 9 builds on this approach along with the earlier theoretical insights to provide a framework for considering amendments to existing wildlife trade policy and regulation, using CITES as a benchmark. It also considers the question of whether the entire approach to wildlife trade regulation should change, as some critics have suggested.

Chapter 10 concludes with a reflection on what the analytical framework teaches us about wildlife trade, limitations to what we have learned, and future avenues for further investigation.

In the examples we consider, readers will notice a bias toward a few well-known case studies, such as the rhino horn and elephant ivory markets. This bias exists partly because we have much experience researching these specific examples and partly because they are widely known and perhaps easier for the reader to identify with. The rhino horn trade example is also especially useful in that it illustrates most of the issues that are typically raised when considering various trade policy options. However, we emphasize that our framework aims to explain broad principles that apply across diverse species and circumstances. We hope that readers will consider how these generalized models might apply to their specific areas of interest within the realm of wildlife trade and conservation.

Notes

1 Non-monetary trade ("barter") entails the direct exchange of goods and services for other goods and services.

2 Some scholars claim that trade, combined with associated specialization, has played a key role in enabling modern human civilization and our relative dominance as a species on planet Earth (e.g., Ridley, 2010).

3 The CITES definition of trade is laid out in Article I of the Convention, and covered by items (c), (d), and (e), which read as follows: (c) "Trade" means export, re-export, import and introduction from the sea; (d) "Re-export" means export of any specimen that has previously been imported; (e) "Introduction from the sea" means transportation into a State of specimens of any species which were taken in the marine environment not under the jurisdiction of any State—see https://cites.org/eng/disc/text.php.

4 The CITES Parties have attempted to address this by way of a CITES Resolution, but the terms remain somewhat open to interpretation—see https://cites.org/sites/default/files/documents/COP/19/resolution/E-Res-05-10-R19.pdf.

5 Some definitions of wildlife trafficking include poaching as a component of it rather than a separate activity.

6 It is beyond the scope of this book to explain concepts such as this, so readers unfamiliar with these terms are encouraged to conduct their own further research.

7 Contemporary versions of this book (e.g., Marshall, 2009) are still available to readers interested in learning more about these concepts.

8 Coase's key insights are distilled in his article "The Problem of Social Cost" (Coase, 1960), which Coase himself claimed was frequently misinterpreted. For a more recent and clear application of his ideas to solving environmental problems, we recommend reading Anderson and Libecap (2014).

9 There are various formal definitions of "institutions" in the literature—one of the most succinct is from Hodgson (2006:18), who defines institutions as "systems of embedded social rules that structure social interactions."

10 Key references in this large body of work include Ostrom (1990, 2005, 2009, 2010) and McGinnis and Ostrom (2014).

11 See, for example, Wandesforde-Smith (2016), Fukushima et al. (2021), Cooney et al. (2021), Wyatt (2021), and Hughes et al. (2023).

12 For earlier discussions using rhinos as illustrative case studies, see 't Sas-Rolfes (2000) and Leader-Williams (2003).

13 A safer way to tackle the problem would have been to engage with consumers first, which conservation scientists have belatedly started to realize (Veríssimo et al., 2012, 2020).

14 For readers who do not find this intuitively obvious, the framework presented in Chapters 5 and 6 can be applied to derive this result. As shown in those chapters, when there are two types of wildlife goods (labeled wild goods and substitute goods in those chapters), subsidizing the substitute goods sellers would lower the price of wildlife goods whether or not consumers can distinguish between the two types of goods; consequently, the supply of the wild goods would decrease. Now, any policy that imposes restrictions on the sellers of the substitute goods and raises their cost of supply has the exact opposite effects of a subsidy—such a policy would thus raise the price of wildlife goods and increase the supply of the wild goods. By relabeling the wild goods as illegal goods and the substitute goods as legal goods in the models presented in Chapters 5 and 6, we obtain the result that restricting the supply of legal rhino horns would raise the price of horns and increase the amount of poaching that supplies the illegally traded horns.

15 Not all social scientists agree that humans make their decisions in this way, an issue we explore further in Chapter 2, in the section on institutions.

16 For a more technical discussion of these concerns, see Fischer (2004).

References

Anderson, T. L., & Libecap, G. D. (2014). *Environmental Markets: A Property Rights Approach*. Cambridge University Press.

Baker, S. E., Cain, R., Kesteren, F. van, Zommers, Z. A., D'Cruze, N., & Macdonald, D. W. (2013). Rough trade: Animal welfare in the global wildlife trade. *BioScience, 63*(12), 928–938. https://doi.org/10.1525/bio.2013.63.12.6

Barnosky, A. D., Koch, P. L., Feranec, R. S., Wing, S. L., & Shabel, A. B. (2004). Assessing the causes of Late Pleistocene extinctions on the continents. *Science, 306*(5693), 70–75.

Becker, G. S. (1968). Crime and punishment: An economic approach. *Journal of Political Economy, 76*(2), 176–177. https://doi.org/10.1086/259394

Blackburn, T. M., Essl, F., Evans, T., Hulme, P. E., Jeschke, J. M., Kühn, I., Kumschick, S., Marková, Z., Mrugała, A., Nentwig, W., Pergl, J., Pyšek, P., Rabitsch, W., Ricciardi, A., Richardson, D. M., Sendek, A., Vilà, M., Wilson, J. R. U., Winter, M., … Bacher, S. (2014). A unified classification of alien species based on the magnitude of their environmental impacts. *PLoS Biol, 12*(5), e1001850. https://doi.org/10.1371/journal.pbio.1001850

Brennan, J. F., & Jaworski, P. (2015). *Markets without Limits: Moral Virtues and Commercial Interests*. Routledge.

Challender, D. W. S., Brockington, D., Hinsley, A., Hoffmann, M., Kolby, J. E., Massé, F., Natusch, D. J. D., Oldfield, T. E. E., Outhwaite, W., 't Sas-Rolfes, M., & Milner-Gulland, E. J. (2022). Mischaracterizing wildlife trade and its impacts may mislead policy processes. *Conservation Letters, 15*(1), e12832. https://doi.org/10.1111/conl.12832

Challender, D. W. S., 't Sas-Rolfes, M., Milner-Gulland, E. J., & Broad, S. (2025). A theory of change to improve conservation outcomes through CITES. *Frontiers in Ecology and Evolution*. https://doi.org/10.3389/fevo.2025.1425267

Coase, R. H. (1960). The problem of social cost. *Journal of Law & Economics, 3* (1), 1–69.

Cooney, R., Challender, D. W. S., Broad, S., Roe, D., & Natusch, D. J. D. (2021). Think before you act: Improving the conservation outcomes of CITES listing decisions. *Frontiers in Ecology and Evolution, 9*. https://doi.org/10.3389/fevo.2021.631556

Di Minin, E., 't Sas-Rolfes, M., Selier, J., Louis, M., & Bradshaw, C. J. A. (2022). Dismantling the poachernomics of the illegal wildlife trade. *Biological Conservation, 265*, 109418. https://doi.org/10.1016/j.biocon.2021.109418

Duffy, R. (2016). War, by conservation. *Geoforum, 69*, 238–248. https://doi.org/10.1016/j.geoforum.2015.09.014

Duffy, R. (2022). *Security and Conservation: The Politics of the Illegal Wildlife Trade*. Yale University Press.

Duffy, R., Massé, F., Smidt, E., Marijnen, E., Büscher, B., Verweijen, J., Ramutsindela, M., Simlai, T., Joanny, L., & Lunstrum, E. (2019). Why we must question the militarisation of conservation. *Biological Conservation, 232*, 66–73. https://doi.org/10.1016/j.biocon.2019.01.013

Ekelund, R. B., & Hébert, R. F. (1983). *A History of Economic Theory and Method* (2nd ed.). McGraw-Hill.

Fischer, C. (2004). The complex interactions of markets for endangered species products. *Journal of Environmental Economics and Management, 48*(2), 926–953. https://doi.org/10.1016/j.jeem.2003.12.003

Fukushima, C. S., Tricorache, P., Toomes, A., Stringham, O. C., Rivera-Téllez, E., Ripple, W. J., Peters, G., Orenstein, R. I., Morcatty, T. Q., Longhorn, S. J., Lee, C., Kumschick, S., de Freitas, M. A., Duffy, R. V., Davies, A., Cheung, H., Cheyne, S. M., Bouhuys, J., Barreiros, J. P., … Cardoso, P. (2021). Challenges and perspectives on tackling illegal or unsustainable wildlife trade. *Biological Conservation, 263*, 109342. https://doi.org/10.1016/j.biocon.2021.109342

Goetschel, A. F. (2024). Animal welfare as the basis of One Health: A UN convention on animal welfare, health, and protection poses a realistic solution to improved animal welfare and human health. *CABI One Health, 3*(1), 1–6. https://doi.org/10.1079/cabionehealth.2024.0003

Harari, Y. N. (2014). *Sapiens: A Brief History of Humankind*. Harvill Secker.

Hardin, G. (1968). The tragedy of the commons. *Science, 162*(3859), 1243–1248.

Hayek, F. A. (1945). The use of knowledge in society. *The American Economic Review, 35*(4), 519–530. www.jstor.org/stable/1809376?origin=JSTOR-pdf

Hodgson, G. M. (2006). What are institutions? *Journal of Economic Issues, 40*(1), 1–25. https://doi.org/10.1080/00213624.2006.11506879

Hübschle, A. (2016). *A Game of Horns: Transnational Flows of Rhino Horn*. IMPRS-SPCE. https://globalinitiative.net/wp-content/uploads/2016/05/Huebschle-A-Game-of-Horns-2016.pdf

Hübschle, A., & Margulies, J. (2024). The need for a socioecological harm reduction approach to reduce illegal wildlife trade. *Conservation Biology, 38*(5), e14335. https://doi.org/10.1111/cobi.14335

Hughes, L. J., Morton, O., Scheffers, B. R., & Edwards, D. P. (2023). The ecological drivers and consequences of wildlife trade. *Biological Reviews, 98*(3), 775–791. https://doi.org/10.1111/brv.12929

Hutton, J., & Webb, G. (2002). Crocodiles: Legal trade snaps back. In S. Oldfield (Ed.), *The Trade in Wildlife: Regulation for Conservation* (illustrated ed., pp. 108–120). Earthscan.

Karesh, W. B., Cook, R. A., Bennett, E. L., & Newcomb, J. (2005). Wildlife trade and global disease emergence. *Emerging Infectious Diseases, 11*(7), 1000–1002. https://doi.org/10.3201/eid1107.050194

Kasterine, A., & Lichtenstein, G. (2018). *Trade in Vicuña: Implications for Conservation and Rural Livelihoods*. International Trade Centre. www.intracen.org/uploadedFiles/intracenorg/Content/Publications/Vicuna_trade_final_Low-res.pdf

Leader-Williams, N. (2003). Regulation and protection: Successes and failures in rhinoceros—Case study. In S. Oldfield (Ed.), *The Trade in Wildlife: Regulation for Conservation* (illustrated edk, pp. 89–99). Earthscan.

Liu, S., Westbury, M. V., Dussex, N., Mitchell, K. J., Sinding, M.-H. S., Heintzman, P. D., Duchêne, D. A., Kapp, J. D., von Seth, J., Heiniger, H., Sánchez-Barreiro, F., Margaryan, A., André-Olsen, R., De Cahsan, B., Meng, G., Yang, C., Chen, L., van der Valk, T., Moodley, Y., … Gilbert, M. T. P. (2021). Ancient and modern genomes unravel the evolutionary history of the rhinoceros family. *Cell, 184*(19), 4874–4885. https://doi.org/10.1016/j.cell.2021.07.032

Marshall, A. (2009). *Principles of Economics: Unabridged Eighth Edition* (8th ed.). Cosimo Classics.

Massé, F., & Margulies, J. D. (2020). The geopolitical ecology of conservation: The emergence of illegal wildlife trade as national security interest and the re-shaping of US foreign conservation assistance. *World Development, 132*, 104958. https://doi.org/10.1016/j.worlddev.2020.104958

McGinnis, M., & Ostrom, E. (2014). Social-ecological system framework: Initial changes and continuing challenges. *Ecology and Society, 19*(2). https://doi.org/10.5751/ES-06387-190230

Milner-Gulland, E. J., & Leader-Williams, N. (1992). A model of incentives for the illegal exploitation of black rhinos and elephants: Poaching pays in Luangwa Valley, Zambia. *Journal of Applied Ecology*, 388–401.

Nellemann, C., Henriksen, R., Raxter, P., Ash, N., & Mrema, E. (2014). The environmental crime crisis: Threats to sustainable development from illegal exploitation and trade in wildlife and forest resources. www.cabdirect.org/cabdirect/abstract/20163106339

North, D. C. (1990). *Institutions, Institutional Change and Economic Performance*. Cambridge University Press.

Norton-Griffiths, M. (1996). Property rights and the marginal wildebeest: An economic analysis of wildlife conservation options in Kenya. *Biodiversity and Conservation, 5*(12), 1557–1577. https://doi.org/10.1007/BF00052116

Norton-Griffiths, M. (2007). How many wildebeest do you need? *World Economics, 8*(2), 41–64.

Ogutu, J. O., Piepho, H.-P., Said, M. Y., Ojwang, G. O., Njino, L. W., Kifugo, S. C., & Wargute, P. W. (2016). Extreme wildlife declines and concurrent increase in livestock numbers in Kenya: What are the causes? *Plos One, 11*(9), e0163249. https://doi.org/10.1371/journal.pone.0163249

Ostrom, E. (1990). *Governing the Commons: The Evolution of Institutions for Collective Action*. Cambridge University Press.

Ostrom, E. (2005). *Understanding Institutional Diversity*. Princeton University Press.

Ostrom, E. (2009). A general framework for analyzing sustainability of social-ecological systems. *Science, 325*(5939), 419–422. https://doi.org/10.1126/science.1172133
Ostrom, E. (2010). Beyond markets and states: Polycentric governance of complex economic systems. *The American Economic Review, 100*(3), 641–672. www.jstor.org/stable/27871226
Ridley, M. (2010). *The Rational Optimist: How Prosperity Evolves*. HarperCollins UK.
Ritchie, H., & Roser, M. (2024). Wild mammals make up only a few percent of the world's mammals. *Our World in Data*. https://ourworldindata.org/wild-mammals-birds-biomass
Roe, D., Dickman, A., Kock, R., Milner-Gulland, E. J., Rihoy, E., & 't Sas-Rolfes, M. (2020). Beyond banning wildlife trade: COVID-19, conservation and development. *World Development, 136*, 105121. https://doi.org/10.1016/j.worlddev.2020.105121
Smith, R. J., & Walpole, M. J. (2005). Should conservationists pay more attention to corruption? *Oryx, 39*(3), 251–256. https://doi.org/10.1017/S0030605305000608
't Sas-Rolfes, M. (2000). Assessing CITES: Four Case Studies. In J. Hutton & B. Dickson (Eds.), *Endangered Species, Threatened Convention* (pp. 69–87). Earthscan.
't Sas-Rolfes, M., Challender, D. W. S., Hinsley, A., Veríssimo, D., & Milner-Gulland, E. J. (2019). Illegal wildlife trade: Scale, processes, and governance. *Annual Review of Environment and Resources, 44*(1), 201–228. https://doi.org/10.1146/annurev-environ-101718-033253
't Sas-Rolfes, M., & Gooden, J. (2024). A conflict of visions: Ideas shaping wildlife trade policy toward African megafauna. *People and Nature, 6*(5), 2029–2045. https://doi.org/10.1002/pan3.10705
't Sas-Rolfes, M., Rademeyer, J., Vigne, L., Emslie, R., Knight, M., Gaymer, J., & Talukdar, B. (2025). The impact of poaching on rhino conservation. In M. Melletti, B. Talukdar, & D. Balfour (Eds.), *Rhinos of the World: Ecology, Conservation and Management*. Springer. https://link.springer.com/book/9783031671685
Taylor, M. S. (2011). Buffalo hunt: International trade and the virtual extinction of the North American bison. *American Economic Review, 101*(7), 3162–3195. https://doi.org/10.1257/aer.101.7.3162
Thomas-Walters, L., Hinsley, A., Bergin, D., Burgess, G., Doughty, H., Eppel, S., MacFarlane, D., Meijer, W., Lee, T. M., Phelps, J., Smith, R. J., Wan, A. K. Y., & Veríssimo, D. (2021). Motivations for the use and consumption of wildlife products. *Conservation Biology, 35*(2), 483–491. https://doi.org/10.1111/cobi.13578
Veríssimo, D., Challender, D. W. S., & Nijman, V. (2012). Wildlife trade in Asia: Start with the consumer. *Asian Journal of Conservation Biology, 1*(2), 49–50.
Veríssimo, D., 't Sas-Rolfes, M., & Glikman, J. A. (2020). Influencing consumer demand is vital for tackling the illegal wildlife trade. *People and Nature, 2*(4), 872–876. https://doi.org/10.1002/pan3.10171
Vigne, L., 't Sas-Rolfes, M., Milliken, T., Rademeyer, J., & Talukdar, B. (2025). The impact of rhino horn trafficking on conservation. In M. Melletti, B. Talukdar, & D. Balfour (Eds.), *Rhinos of the World: Ecology, Conservation and Management*. Springer. https://link.springer.com/book/9783031671685
Wandesforde-Smith, G. (2016). Looking for law in all the wrong places? Dying elephants, evolving treaties, and empty threats. *Journal of International Wildlife Law & Policy, 19*(4), 365–381. https://doi.org/10.1080/13880292.2016.1248701
Wilson, B. J. (2020). *The Property Species: Mine, Yours, and the Human Mind*. Oxford University Press.

Wilson, B. J. (2025). *Meaningful Economics: Making the Science of Prosperity More Human*. Oxford University Press.
Wolff, A. Wm. (Ed.). (2023). The origins of trade. In *Revitalizing the World Trading System* (pp. 9–18). Cambridge University Press. https://doi.org/10.1017/9781009289290.004
Wyatt, T. (2021). *Is CITES Protecting Wildlife?: Assessing Implementation and Compliance* (1st ed.). Routledge.

2 Wildlife Conservation and Trade

Theory and Practice

Before taking a deeper dive into our economic analysis of wildlife markets, we consider it important to provide some contextual information on three aspects. First, we take a more detailed look at the term "wildlife conservation," which means different things to different people, resulting in some disagreement over the best way to conserve wildlife in today's world. We should consider the source and nature of such disagreement when considering wildlife trade's role and practice. Second, we discuss the role of institutions in shaping human behavior (including that in relation to both wildlife conservation and trade), and we explain some key concepts. Third, we focus more specifically on wildlife trade regulation aimed at supporting conservation. For all three aspects, we cover both theoretical and practical issues, which help set the scene for our subsequent chapters.

Conservation in the Anthropocene

Most scientists today agree that, due to the recent historical combination of unprecedented human population expansion and associated technological advancement and resource use, our species is profoundly influencing the planet's natural systems. Our impact is so great at the global scale that many agree that we have entered a new geological era termed the "Anthropocene" (Steffen et al., 2007; Reyers et al., 2018) and that this has significant implications for how we should approach environmental management and governance, including what we conventionally refer to as "wildlife conservation" (Rockström et al., 2024; Lorimer, 2015). Until the nineteenth century, the term "conservation" applied to human creations (e.g., art and architecture), and it has only been extended to elements of the natural environment in recent history. The use of the term has also evolved and, as with our brief exploration of economic history in Chapter 1, examining this historical evolution of conservation ethics, policy, and practice provides some lessons.

DOI: 10.4324/9781003286318-2

Conservation Ethics

Although ethical concern for wildlife and the natural environment has worldwide cultural roots that may extend far back in time, the formal wildlife conservation movement first gained traction in the late nineteenth and early twentieth centuries, heavily influenced by the tension between two different philosophical approaches in the USA (Callicot, 1994; Dyke & Lamb, 2020). One approach, grounded in more spiritual notions of the value of "wilderness" and popularized by John Muir, advocated the "preservation" of natural areas with minimal human presence and intervention and inspired the creation of state-protected areas such as National Parks (Nash & Miller, 2014). The other, grounded in the practice of forestry and the principle of "wise" or "sustainable" use of natural resources, gave rise to the alternative term "conservation" promoted by Gifford Pinchot in the early twentieth century and underpinned a more utilitarian view of wildlife management. In the mid-twentieth century, these contrasting approaches were somewhat reconciled by Aldo Leopold, who promoted a "land ethic" as a more holistic ecological approach to managing wildlife in natural areas (Callicot, 1990).

While conservation started as a social and political movement informed by strands of knowledge from forestry, agriculture, and biological science, a unified discipline of "conservation biology" started to emerge in the late 1970s (Jepson & Ladle, 2010). In the 1980s, it was established formally through the creation of the Society for Conservation Biology (SCB) and was defined as a "crisis discipline" grounded in both scientific inquiry and an ethical imperative to conserve biological diversity, ecological complexity, and evolutionary processes (Soulé, 1985). The 1970s also saw the ascendance of concern for the welfare of individual animals, inspired by the writing of philosopher Peter Singer (1975) and growing public consciousness around the sentience of animal species such as great apes, elephants, and whales. This reinvigorated an older animal protection movement (previously focused on domestic animals), introduced the notion of "animal rights" into the discourse, and eventually engendered a "compassionate conservation" movement (Bekoff, 2013), which places high value on the welfare of individual wild animals relative to collective entities such as species, populations, and ecosystems.[1]

Environmental ethicists and others who support the SCB and compassionate conservation approaches promote the notion that nature and its components (especially individual wild animals) possess "intrinsic value" that is "non-anthropocentric" (Batavia & Nelson, 2017). Intrinsic value is distinguished from anthropocentric "instrumental" values, which provide utility to humans. Philosophers argue over whether the notion of intrinsic value is practically relevant (Callicot, 2002), but the existence of this debate indicates the extent of disagreement in approaches among environmentalists.[2] Some members of the conservation biology community have realigned their approach to one of "conservation science," being differentiated by a view that humans dominate the world and should be placed at the center of effective strategic approaches to conservation (Karieva & Marvier, 2012).

Related to the broader debate over whether conservation should be human-centered or not, in the late twentieth century, the conservation community proposed numerous other normative concepts, most of which can be loosely assigned to one of two contrasting categories (Callicot et al., 1999). The first "compositionalist" category is more conventionally focused on organisms and concerned with species composition in ecosystems, whereas the second "functionalist" category is more concerned with the biophysical processes of ecosystem function such as nutrient cycles. The categories are not mutually exclusive, but each has more relevance in different contexts across varying scales.

In the early twenty-first century, conservationists have increasingly called for integrating opposing ethical viewpoints and adopting an integrated "people and nature" or "stewardship approach" (Tallis & Lubchenco, 2014; Mace, 2014; Weddell, 2023). However, even approaches to integrating people vary (Holmes et al., 2017). For example, one approach emphasizes strengthening property rights and developing appropriate market institutions (Anderson & Libecap, 2014); another, termed "convivial conservation" (Büscher & Fletcher, 2019), is founded in opposition to capitalism. A third falls within the broader social-ecological systems-oriented approach of "sustainability science" and emphasizes concepts such as system resilience and adaptive polycentric governance (Norton, 2005; Folke, 2007; Folke et al., 2016). Despite all attempts to integrate varying viewpoints, conservation decision-makers will continue to face complicated trade-offs between, for example, the interests of individual organisms (including humans) and various collective entities, and there appears to be no simple universal ethical theory that can be applied to resolve these.

Policy and Practice

The reality of the trade-offs that conservationists face becomes apparent in the world of policy and practice. International conservation policy has evolved significantly since the early twentieth century and continues to do so, influenced by human development, biophysical effects of climate change, shifting geopolitics, and advances in scientific knowledge. While early formal conservation efforts were driven by civil collectives in North America and Europe with limited national government support, from the mid-twentieth century, governments and intergovernmental agencies started to play a greater role. This was facilitated by the establishment of the International Union for the Conservation of Nature (IUCN), which was instrumental in drafting CITES (Boardman, 1981). The emergent composition of the IUCN reflects a range of interests, including protected areas, species conservation, ecosystem management, and climate change, as well as a plethora of approaches to address specific issues.

In 1972, the United Nations convened a Conference on the Human Environment, which spawned the United Nations Environment Programme (UNEP), a multilateral agency that facilitates intergovernmental cooperation to work on a range of interconnected environmental issues. In 1980, the IUCN collaborated with UNEP and other organizations to produce the World Conservation Strategy, a document that stated three central objectives of conservation: (1) the maintenance of essential ecological processes and life-support systems, (2) the preservation of genetic diversity, and (3) the sustainable utilization of species and ecosystems. During the 1980s, as the concept of biological diversity gained traction and tended to replace the word "wildlife" in policy discourse, "biodiversity" (Wilson, 1988) became popularized as a central focus for conservation policy.

In 1992, the United Nations Conference on Environment and Development (held in Rio de Janeiro) spawned another conservation treaty, the Convention on Biological Diversity (CBD), which rapidly gained wider support than CITES (Escobar-Pemberthy & Ivanova, 2020) and promotes a broader, more inclusive approach to wildlife conservation, consistent with the over-arching goal of sustainable development. The CBD includes protocols on biosafety and on access and benefit sharing in relation to genetic resources. In 2015, the United Nations Member States adopted 17 Sustainable Development Goals (SDGs), which include ending poverty, reducing inequality, and building more peaceful, prosperous societies by 2030. The goals also address climate action, life on land and below water, and responsible consumption and production. In alignment with the SDGs, in 2022, the Parties to the CBD agreed to the Kunming-Montreal Global Biodiversity Framework, which includes an ambitious and wide-ranging list of 23 global targets to be implemented by 2030. Significantly, all these targets have a strong anthropocentric orientation, with one possible exception being Target 3, which is to conserve 30% of the earth's land, waters, and seas. However, Target 3 also provides for sustainable use that "is fully consistent with conservation outcomes."

Government policies toward wildlife conservation tend to be more people-oriented than those of nongovernmental organizations with more specific aims (such as animal protection). Since governments have ultimate jurisdiction over most areas where wildlife occurs, they must create relevant institutions (e.g., pass legislation and establish agencies and regulations) and raise sufficient finance for effective implementation. Government agencies responsible for conservation face numerous practical challenges, with adequate funding typically at the top of the list, especially in developing countries. Apart from maintaining protected areas, government agencies must also manage many individual wild species to mitigate problems such as human–wildlife conflict and ecosystem degradation caused by alien invasives or overpopulation. In many countries, wildlife agencies address the dual challenges of raising finance and managing species through selective trade of harvesting rights (e.g., selling hunting and fishing licenses[3] and enabling sustainable plant harvesting), sometimes also allocating such rights to local communities (Child, 2019).

As another strategy, conservation practitioners (state and private) frequently manage species in controlled environments. "*Ex situ*" species conservation refers to individuals removed from their natural habitats and protected in intensively managed conditions with artificial support (e.g., zoos and plant nurseries) and is contrasted with "*in situ*" conservation. However, species are increasingly managed along a continuum between the two categories, often for partly commercial or economic reasons. There are concerns that overmanaged and manipulated species will evolve differently and unnaturally compared to those that experience "natural" selection pressures in the wild, prompting conservation scientists to analyze this question carefully in the case of herbivores on wildlife ranches in South Africa (Child et al., 2019). Potential species modification or extinction via domestication thus presents another interesting challenge to conservationists. Nevertheless, the multiple growing resource pressures of the Anthropocene dictate that future wildlife conservation efforts must embrace some pragmatism in their approach (Marris, 2011).

Institutions Matter

As mentioned in Chapter 1, economists and other social scientists have identified institutional arrangements as critical determinants of both economic performance and conservation outcomes. Institutions are more than simply stated rules—they are rules that people are motivated to (and actually do) follow (Guala, 2016). They may be informal (e.g., customs and established social norms) or formal (e.g., written laws enforced by governments), and they both constrain some human behaviors and enable others (North, 1991; Lauth, 2015).[4] Concerning wildlife conservation and trade, the types of institutions of greatest interest to us are (1) laws that regulate ownership, access to, and use of wildlife (which we can mostly interpret as property rights or constraints on them), (2) rules that regulate trading activity, and (3) money.

Within the social sciences, there is some disagreement over how and why institutions shape human behavior. These disagreements are grounded in differing characterizations of human nature. One school of thought views humans as essentially individualistic and rational, with their actions largely dictated by assessments of benefits and costs. Another school emphasizes that humans are social and cultural beings, strongly influenced by a need to conform or belong, and therefore should be studied collectively. To some extent, these disagreements affect the influence of standard neoclassical economic analysis, which is mostly grounded in the rational individualism school. However, the broader discipline of economics is shifting to embrace new insights gained from psychology and experimental economic research, which have given rise to the field of behavioral economics (Kahneman, 2013; Thaler, 2016) and a more integrated "humanomics" approach (Smith & Wilson,

2019). Scholars of environmental governance view the varying schools of social science as largely complementary (e.g., Ostrom, 1990), and standard economic analysis remains a compelling way of analyzing human behavior in certain contexts if we think of humans applying "bounded rationality" to their decision-making (Simon, 1955).

Institutions have featured prominently in the literature on natural resource economics, which integrates institutional insights into "bioeconomic" models to determine the optimal harvest rates of wild species. The seminal bioeconomics literature mainly examined forestry and fisheries ('t Sas-Rolfes et al., 2019), and this may have inappropriately influenced early regulatory approaches toward terrestrial animals (including the design of CITES), as such regulation should take greater account of land conversion pressures and habitat management costs (Swanson, 1994). More recent literature that applies economic models to managing biodiversity as a resource deals specifically with terrestrial animals (e.g., Milner-Gulland & Mace, 1998; van Kooten & Bulte, 2000; Kontoleon et al., 2007) but varies on how it considers institutional aspects in relation to wildlife harvesting and trade.

Property Rights

Returning to the American bison example from Chapter 1, although the trade-induced exploitation spike of the 1870s was a critical factor in population depletion, several other factors were involved over the longer term, including the role of property rights regimes (Lueck, 2002). Nineteenth-century bison were effectively open-access resources, and the species' subsequent recovery has occurred in the context of "stronger" property rights.[5] Unlike most other wildlife species in the USA, some states reclassified bison as agricultural animals, enabling private ownership and trade in their meat and hides, and bison populations have since recovered on both private land (where they are harvested) and in state parks (where they are not).[6] Other large wild animals in the USA are managed under the North American model of wildlife management (Geist, 1994; Mahoney & Geist, 2019), which rejects private wildlife ownership, places wildlife under "public trust" (i.e., state ownership and control), and prohibits commercial trade in wildlife products. Other countries, notably in Southern Africa, have adopted an approach that more closely aligns with the bison recovery—i.e., they have enabled at least some extent of wildlife "ownership" or control by non-state actors, and allowed and even enabled legal wildlife markets (Child, 2019).

The approach adopted in Southern Africa aligns with economic theories of property rights, which rest on two propositions. The first is that outcomes will be determined by property rights regimes—i.e., the best resource management performance will align with the most suitable allocation of property rights for a particular good and context. Examples of different regimes are private property, common property, state property, and open access (Bromley, 1989). The second proposition is that the most suitable allocation will be determined by the inherent economic nature of the good in question, according to two characteristics: rivalry in consumption and the ability to exclude (Ostrom, 2010). The possible combinations of these two characteristics yield four categories: private goods, toll goods, public goods, and common pool resources. Noting that wildlife species may yield multiple products (also linked to their habitats), their inherent nature overall might comprise a mix of these characteristics (Lueck & Miceli, 2007). An empirical study on rhino conservation performance in Africa supported these propositions, demonstrating that over time, rhinos survived best in countries that enabled the property rights regimes that most closely matched their inherent economic characteristics ('t Sas-Rolfes & Emslie, 2024).

Economists have also studied the evolution of property rights and observed that when people see the potential to capture personal benefits from an unclaimed resource (including wildlife), they typically try to create and acquire private property rights (Anderson & Hill, 1975; Lueck & Torrens, 2020). It is ultimately up to governments to formalize, allocate, and enforce these rights. Property rights allocations determine the distributions and flows of benefits and costs. In many instances, a property regime may be misaligned with the economic nature of a good, resulting in some benefits or costs spilling over to parties other than the owner (e.g., pollution), creating so-called "externalities" and raising the question of whether government intervention could help address them through, for example, regulations, taxes, or subsidies. In his influential article, "The Problem of Social Cost," Ronald Coase (1960) demonstrated that property rights allocations matter because of how they establish and distribute the consequent transaction costs of solving these problems, whether by government intervention or private contracting. Property rights allocations thus have implications for both economically fair distribution and efficient use of resources. Relevant to conservation, they will strongly affect the incentives to harvest wildlife at sustainable rates (rather than to extinction).

Market Institutions and Transaction Costs

In society, entrepreneurs[7] seek creative opportunities to make themselves better off, usually by benefitting others in the process.[8] In some instances, they can achieve this by radically modifying property rights regimes,[9] but doing that typically involves much time and effort. Therefore, in most cases, they do so by identifying trading opportunities and ways to reduce costs, including transaction costs. In a theoretical "free market," the role of government would be restricted to defining and enforcing property rights. In practice, citizens influence governments to shape market activity in various ways, either by directly modifying property rights[10] or indirectly, through various forms of regulation (e.g., laws and standards) or market intervention (e.g., direct participation, price controls, setting quotas, taxes, and subsidies). Such regulation or market intervention may also affect the strength of property rights and, therefore, the potential tradable value of goods.

In most countries, wildlife conservation is a heavily regulated sector, and governments typically suppress market activity.[11] However, there are numerous wildlife goods and services for which demand exists, and entrepreneurs strive to meet these demands, either within or outside the law, through contracting and trading. Market participants who do not accept the ethical premises of government regulations will simply view these as transaction costs to overcome. Trading activity may take place through barter but typically involves money, which is used to determine prices. Understanding how prices work in relation to scarcity and value is crucial to understanding markets, including those that are not readily obvious.

As mentioned in the previous chapter, prices convey information about scarcity in the marketplace. Note that they do this in relative terms: a single price data point does not tell us much, but we learn a lot when we compare it to other data points. Such other data points could include prices for the same product at a different time or place, prices paid for other products, and how much money we (and others) have available to spend. These comparisons tell us a lot about the circumstantial scarcity of whatever was traded and how that might change over time and space. Also, note that the price of something is not an objective reflection of absolute value (nor can it be); it is a more subjective reflection of relative value.

In a theoretical, sufficiently competitive market, prices most accurately convey a good or service's relative social market value. In practice, markets are typically influenced by factors that alter the price and thereby mask the social market value. Distorting factors include the various government interventions in markets as described above (e.g., price controls and taxes). Another common distorting factor is monopoly power, which comes about when either a buyer or a seller faces limited to no competition, which enables the seller (/buyer) to negotiate a higher (/lower) price than would otherwise be possible. In the case of monopoly sellers, they have incentives to both raise prices and reduce their supply, which may be considered positive when the objective is to reduce the output of something like a wildlife good.[12]

So, how do market institutions influence wildlife extraction rates and, therefore, conservation? Given an obvious demand for conserving wildlife *in situ* and consuming wildlife products obtained from extraction, theoretical markets exist for both. In practice, most markets in the wildlife conservation and product sectors are heavily regulated and severely distorted. However, to the extent that governments enable the establishment of functional property rights (i.e., strong and appropriately allocated) and refrain from distorting the market through intervention, trading regimes will determine the desired extent of conservation and extraction through the pricing system. Conservationists are prone to being pessimistic about how such market processes would play out, often assuming they would result in species extinction. However, this pessimism is largely unfounded if markets are appropriately designed to capture and reinvest the value from harvesting into conservation.

Consider the example of rhino conservation in South Africa. For several decades, the South African government has allowed domestic trade in both live rhinos and rhino hunting experiences, providing a useful data set to compare the value of live rhinos (as revealed by auction prices) with that of dead rhinos (as revealed by hunting fees). Because South Africa also provides a reasonably accurate data set of rhino numbers over this time, we can assess the impact of the extraction through hunting as well as that of other policy measures. The data demonstrate that extraction through hunting did not negatively impact the southern white rhino species—quite the opposite, as the income from hunting provided a strong incentive to conserve more rhinos ('t Sas-Rolfes et al., 2022). The price data also show that white rhinos went from being worth far more dead than alive to the reverse as property rights were strengthened and markets for live rhinos enabled[13] ('t Sas-Rolfes & Emslie, 2024). The rhino example demonstrates how we must carefully examine the interacting influences of property rights, prices, and government policies that impact wildlife markets. It is these interacting dynamics that will determine whether wildlife extraction rates are sustainable or not—the relationship between these factors and recorded trade activity is complicated.

Wildlife Trade Regulation

Having discussed the complexity of conservation goals and the significance of institutions, we now consider the role of wildlife trade regulation in affecting conservation outcomes. As mentioned in Chapter 1, meaningful attempts to regulate cross-border wildlife trade (instead of just hunting or harvesting) can be traced back to the USA and colonial Europe in the early twentieth century. However, the establishment of CITES in 1975[14] embedded wildlife trade regulation as a mainstream global approach to wildlife conservation. Initially claimed as a success (Lyster, 1985), this view soon started to be questioned (Kosloff & Trexler, 1987), and after five decades, the treaty has received much criticism from various quarters for varying reasons ('t Sas-Rolfes et al., 2025; Challender et al., 2025). Although some of the identified problems are political in nature, many of the shortcomings appear to arise from a lack of alignment between the CITES approach and the underlying economics of wildlife trade.

Theory

Acknowledging a rich literature on regulatory theory (e.g., Drahos, 2017), we can identify at least two distinct approaches to wildlife trade regulation: a more conventional legalistic approach and a more economic approach. In essence, the legalistic approach uses technical input from scientists to determine desirable objectives and then creates laws to achieve them directly, assuming that most people will simply adopt and comply with those laws and that laws influence social norms (Chambers, 2008). The economic approach is more circumspect and evaluates all proposed laws in terms of the probability and costs of effective implementation, enforcement, and compliance. The economic approach examines institutional incentives, being conscious that laws may conflict with social norms and thus face resistance.[15] When such conflict occurs, laws may have unintended consequences, such as eliciting responses that cause perverse outcomes (that are inferior to the desired objective and possibly even worse than the original situation).

The legalistic approach that underpins CITES follows a relatively simplistic logic. Species deemed "threatened by trade" by a country's appointed scientific authority become candidates for listing. Depending on the assessed level of threat, the species may be listed in one of three Appendices. Appendix I listings are the most tightly regulated, amounting to a prohibition on all international commercial trade. Appendix II listings allow trade within a system of import and export permits. Appendix III listings relate to species that are legally protected within their native range in particular countries, and for which such countries request international cooperation to take action against illegal trade in other countries. Species may be uplisted or downlisted, with the former happening far more frequently than the latter. CITES prescribes that all listing decisions (up or down) should be guided by the "precautionary principle" (Kriebel et al., 2001), but its exact application to wildlife trade policy is somewhat unclear (Dickson, 1999).

An economic approach to wildlife trade regulation, which has been proposed[16] but largely ignored by the CITES Parties, would consider both broader environmental objectives beyond species conservation (e.g., climate change solutions, habitat conservation, and other ecosystem functioning concerns) and the broader institutional context (such as the alignment of formal and informal institutions). As a starting point, this approach would tackle unsustainable levels of species extraction by trying to increase the perceived *in situ* value of living populations of the species to the local people who have the most control over their destiny. Simplistic attempts to deter extraction through prohibition and (often violent) enforcement may fail to achieve this objective in the long term (Duffy et al., 2019).

Following proposals to adopt a more economic approach toward elephant conservation and the ivory trade (Barbier et al., 1990), the literature contains an extensive (largely theoretical) debate over ivory trade regulation and potential elephant conservation consequences and, to a lesser extent, debates about rhino horn and tiger product trades ('t Sas-Rolfes et al., 2019). This literature demonstrates that trade regulation is highly complicated and that specificity must be considered to achieve conservation goals. The most appropriate approach will differ between species and even within species, between different locations (affected by geography and varying institutions). Furthermore, trade regulation must consider that conservation goals and economic (livelihood) goals are interconnected, which implies that factors must also be considered in the light of potential feedback effects and that both sets of goals should be considered simultaneously (Cooney et al., 2015).

An especially challenging issue for wildlife trade regulation concerns the growing trend toward commercial captive breeding and "farming" of wild species. The literature on this topic remains somewhat naïve concerning the dynamic interactions between those trading wild products and those trading substitute "farmed" products and how this should be regulated (Hinsley & 't Sas-Rolfes, 2020). While there are many cases in which farmed products may provide acceptable and competitive alternatives to wild products (e.g., crocodiles, ostriches, and many plant species), in other instances the role of farming may be more ambiguous for conservation (Bulte & Damania, 2005; 't Sas-Rolfes & Challender, 2020) and may require innovative institutional interventions to ensure that any permitted farming subsidizes or otherwise supports the *in situ* conservation of the species (Abbott & van Kooten, 2011). We explore the theoretical aspects of this issue further in Chapters 5 and 6 of this book while also considering similar concerns about the effect of introducing competing substitute goods created through synthetic biology or other technologies.

Practice

To date, with some exceptions, the practical experience of wildlife trade regulation has been somewhat unimpressive in terms of verifiable conservation achievements, and it has drawn much criticism. CITES has evolved significantly since its inception[17] (Wijnstekers, 2018), with periodic reviews of the effectiveness of its approach to achieving conservation (Hepworth, 1998; Oldfield, 2003). Supporters of the legalistic approach to regulation have complained about the Parties' inadequate implementation, which they blame on insufficient resourcing due to a lack of political will (e.g., Reeve, 2002; Wyatt, 2021). Others have questioned the rationale of the legalistic approach (e.g., Hutton & Dickson, 2000; Challender & MacMillan, 2014). Recent research identifies varying motivations and incompatible agendas within the wildlife trade regulation community and finds that the structure of CITES entrenches systemic policy conflict ('t Sas-Rolfes & Gooden, 2024; 't Sas-Rolfes et al., 2025).

The creation of CITES was followed by the establishment of nongovernmental organizations focused specifically on wildlife trade, the first of which was TRAFFIC, which is affiliated with the IUCN. Over the years, TRAFFIC has conducted extensive research on wildlife trade, assisted governments with compliance and enforcement matters, and played both an informational and policy advocacy role at CITES meetings. Numerous other nongovernmental organizations engage with CITES and have played a strongly influential role (Princen, 1994; Challender & MacMillan, 2019), causing past meeting agendas to be disproportionately dominated by concerns for charismatic species of public interest, notably elephants, rhinos, and big cats.

However, in recent years, public interest has increasingly extended to other species (e.g., pangolins, sharks, and turtles). In recent decades, there has also been a growing number of nongovernmental organizations that are more intensely focused on animal welfare and trade prohibition support (through, for example, undercover work and consumer demand reduction advocacy).

Three issues stand out when tracking CITES progress and its regulatory approach over the last five decades. The first is that, as outlined in the previous chapter, many conservationists, including scientists, remain confused about how CITES works and have unsubstantiated opinions about the role of trade in influencing conversation outcomes. A recent review found that wildlife "trade" is frequently mis-specified, even in the peer-reviewed literature, and that researchers often misunderstand and misinterpret CITES trade data (Challender et al., 2022). A systematic review of the impact of trade restrictions on key species in Southern Africa revealed that the causal relationships between these and conservation outcomes remain poorly researched and understood, with limited evidence of positive long-term results (Hiller & 't Sas-Rolfes, 2025). Experience has also shown that when the relationship between trade and conservation outcomes is uncertain, the CITES Parties have (inappropriately) tended to interpret the precautionary principle to favor trade restrictions (Trouwborst et al., 2020). An unfortunate potential consequence is that "trade" may be diagnosed as a threat to a species when other factors are more relevant (e.g., habitat loss through land conversion), and there may be instances in which restricting trade aggravates those other factors (Cooney et al., 2021).

The second issue of concern is the extent to which the treaty appears to be increasingly imposing regulatory burdens on Parties and simultaneously criminalizing wild species trade, thereby expanding the remit of illegal activity without commensurate increases in resources to combat wildlife crime. By regularly adding more species to the Appendices, the treaty is placing ever-increasing demands on under-resourced governments to implement and enforce the consequent regulations. Illegal wildlife trade (both poaching and trafficking) has proliferated and raised the stakes for environmental governance (Felbab-Brown, 2017). Conventional law enforcement efforts have been challenged by new technological developments (e.g., a dramatic increase in online trading) and the increased role of sophisticated and organized crime syndicates with experience in international trade logistics (involving other products, including drugs and arms). Reviewing recent experience with the illegal elephant ivory and rhino horn trades highlights the potential for path-dependent feedback loops, whereby continuing product scarcity underpins persistently high prices for the illegal products, professionalized wildlife crime systems become deeply entrenched, and the subject species become trapped in a "poaching pit" (Di Minin et al., 2022).

The final issue is that analysis of the long-term CITES trade data reveals that wildlife is increasingly being farmed for commercial purposes (Harfoot et al., 2018), likely because the treaty is more lenient toward trade in species produced in captivity. Numerous challenges are associated with this trend (Meeks et al., 2024). In the case of animals, it raises welfare concerns and associated increases in disease risk. In some instances, farming operations may source wild-extracted specimens for restocking or otherwise launder illegally extracted wild products, enabling them to be traded legally. There are also concerns that providing farmed products to the market has the long-term effect of entrenching the social acceptability of wildlife products, possibly resulting in eventual increased demand for wild-harvested products, which may be considered superior. There is empirical evidence to support these concerns in certain cases, but there are also instances in which wildlife farming can provide *in situ* conservation benefits to species, and it is thus important to discern the factors that determine success or failure, as we discuss in Chapters 5 and 6.

Notes

1. Note that the ideals of compassionate conservation signal a radical departure from the understanding of conservation as the "sustainable use of biological resources"—it views nonhuman animals as sentient beings with their own interests rather than mere "resources" for human use.
2. It also has implications for policy and the way in which conservationists may view economics (which is widely perceived to only consider instrumental values)—see Spash and Hache (2022).
3. This accounts for a significant portion of revenue for most individual state conservation agencies in the USA.
4. Human languages are base institutions upon which all others are built (Searle, 2005).
5. The strength of a property right is determined by how well it is defined and enforced, as well as its durability and transferability (Alchian & Demsetz, 1973).
6. There are mixed opinions on the bison recovery in the USA. It is celebrated by some but somewhat criticized by others, who argue that genetically pure, wild, and unmanaged bison populations remain limited (Freese et al., 2007).
7. Note that entrepreneurs can exist in all sectors of society, including within governments.
8. Although some entrepreneurs may act in a purely selfish way, society typically frowns on such behavior and so they are more likely to succeed when also creating benefits for others.
9. An example of this would be when, in the late 1980s, South African citizens lobbied for new legislation to recognize private ownership of wildlife on fenced conservancies. The resulting Game Theft Act of 1991 unlocked substantial economic value in South Africa's wildlife sector, leading to impressive increases in both privately conserved habitat and numbers of many wild species (Taylor et al., 2020).
10. Note that in most countries that operate as functioning constitutional democracies, it is difficult to fundamentally modify or reallocate property rights because existing right holders typically enjoy strong legal protection against confiscation of their property, even by the government.

11 Nongovernmental organizations in the wildlife sector also tend to avoid or actively oppose market approaches to conservation and lobby governments accordingly.
12 Note that this is only true if the monopoly has not been established in a way that is still subject to strong competition from illegal suppliers, as this could stimulate poaching.
13 This was set back when the South African government subsequently overreached with domestic regulations, raising the effective costs of conserving live animals and thereby reducing the incentives to do so.
14 The treaty was signed in 1973 and came into effect in July 1975 after being ratified by sufficient Parties.
15 As an example of the economic approach, "smart regulation" advocates deeper engagement with civil society at all stages of the regulatory process (Gunningham & Sinclair, 2017).
16 See Bulte et al. (2003)—available at https://cites.org/sites/default/files/eng/prog/economics/CITES-draft6-final.pdf
17 It has also grown—by early 2025, there were 185 CITES Parties and close to 41,000 listed species.

References

Abbott, B., & van Kooten, G. C. (2011). Can domestication of wildlife lead to conservation? The economics of tiger farming in China. *Ecological Economics, 70*(4), 721–728. https://doi.org/10.1016/j.ecolecon.2010.11.006

Alchian, A. A., & Demsetz, H. (1973). The property right paradigm. *The Journal of Economic History, 33*(1), 16–27.

Anderson, T. L., & Hill, P. J. (1975). The evolution of property rights: A study of the American west. *The Journal of Law & Economics, 18*(1), 163–179.

Anderson, T. L., & Libecap, G. D. (2014). *Environmental Markets: A Property Rights Approach*. Cambridge University Press.

Barbier, E. B., Burgess, J. C., Swanson, T. M., & Pearce, D. W. (1990). *Elephants, Economics and Ivory*. Earthscan.

Batavia, C., & Nelson, M. P. (2017). For goodness sake! What is intrinsic value and why should we care? *Biological Conservation, 209*, 366–376. https://doi.org/10.1016/j.biocon.2017.03.003

Bekoff, M. (Ed.). (2013). *Ignoring Nature No More: The Case for Compassionate Conservation*. University Of Chicago Press.

Boardman, R. (1981). *International Organization and the Conservation of Nature*. Macmillan.

Bromley, D. W. (1989). Property relations and economic development: The other land reform. *World Development, 17*(6), 867–877. https://doi.org/10.1016/0305-750X(89)90008-9

Bulte, E. H., & Damania, R. (2005). An economic assessment of wildlife farming and conservation. *Conservation Biology, 19*(4), 1222–1233. https://doi.org/10.1111/j.1523-1739.2005.00149.x

Bulte, E. H., Van Kooten, G. C., & Swanson, T. (2003). *Economic Incentives and Wildlife Conservation*. https://cites.org/sites/default/files/eng/prog/economics/CITES-draft6-final.pdf

Büscher, B., & Fletcher, R. (2019). Towards convivial conservation. *Conservation and Society, 17*(3), 283. https://doi.org/10.4103/cs.cs_19_75

Callicott, J. B. (1990). Whither conservation ethics? *Conservation Biology, 4*(1), 15–20. https://doi.org/10.1111/j.1523-1739.1990.tb00261.x
Callicott, J. B. (1994). *Earth's Insights: A Multicultural Survey of Ecological Ethics from the Mediterranean Basin to the Australian Outback*. University of California Press.
Callicott, J. B. (2002). The pragmatic power and promise of theoretical environmental ethics: Forging a new discourse. *Environmental Values, 11*(1), 3–25. https://doi.org/10.3197/096327102129340957
Callicott, J. B., Crowder, L. B., & Mumford, K. (1999). Current normative concepts in conservation. *Conservation Biology, 13*(1), 22–35. https://doi.org/10.1046/j.1523-1739.1999.97333.x
Challender, D. W. S., Brockington, D., Hinsley, A., Hoffmann, M., Kolby, J. E., Massé, F., Natusch, D. J. D., Oldfield, T. E. E., Outhwaite, W., 't Sas-Rolfes, M., & Milner-Gulland, E. J. (2022). Mischaracterizing wildlife trade and its impacts may mislead policy processes. *Conservation Letters, 15*(1), e12832. https://doi.org/10.1111/conl.12832
Challender, D. W. S., & MacMillan, D. C. (2014). Poaching is more than an enforcement problem. *Conservation Letters, 7*(5), 484–494. https://doi.org/10.1111/conl.12082
Challender, D. W. S., & MacMillan, D. C. (2019). Investigating the influence of non-state actors on amendments to the CITES appendices. *Journal of International Wildlife Law & Policy, 22*(2), 90–114. https://doi.org/10.1080/13880292.2019.1638549
Challender, D. W. S., 't Sas-Rolfes, M., Milner-Gulland, E. J., & Broad, S. (2025). A theory of change to improve conservation outcomes through CITES. *Frontiers in Ecology and Evolution*. https://doi.org/10.3389/fevo.2025.1425267
Chambers, W. B. (2008). *Interlinkages and the Effectiveness of Multilateral Environmental Agreements*. United Nations University Press.
Child, B. (2019). *Sustainable Governance of Wildlife and Community-Based Natural Resource Management* (1st ed.). Routledge.
Child, M. F., Selier, S. A. J., Radloff, F. G. T., Taylor, W. A., Hoffmann, M., Nel, L., Power, R. J., Birss, C., Okes, N. C., Peel, M. J., Mallon, D., & Davies-Mostert, H. (2019). A framework to measure the wildness of managed large vertebrate populations. *Conservation Biology, 33*(5), 1106–1119. https://doi.org/10.1111/cobi.13299
Coase, R. H. (1960). The problem of social cost. *Journal of Law & Economics, 3*, 1.
Cooney, R., Challender, D. W. S., Broad, S., Roe, D., & Natusch, D. J. D. (2021). Think before you act: Improving the conservation outcomes of CITES listing decisions. *Frontiers in Ecology and Evolution, 9*. https://doi.org/10.3389/fevo.2021.631556
Cooney, R., Kasterine, A., MacMillan, D., Milledge, S., Nossal, K., Roe, D., & 't Sas-Rolfes, M. (2015). *The Trade in Wildlife: A Framework to Improve Biodiversity and Livelihood Outcomes*. International Trade Centre. www.intracen.org/uploadedFiles/intracenorg/Content/Publications/2014-2015-76_Low%20Res%20PDF_Trade%20in%20Wildlife(4).pdf
Di Minin, E., 't Sas-Rolfes, M., Selier, J., Louis, M., & Bradshaw, C. J. A. (2022). Dismantling the poachernomics of the illegal wildlife trade. *Biological Conservation, 265*, 109418. https://doi.org/10.1016/j.biocon.2021.109418
Dickson, B. (1999). The precautionary principle in CITES: A critical assessment. *Natural Resources Journal, 39*(2), 211–228. JSTOR.
Drahos, P. (2017). *Regulatory Theory: Foundations and Applications*. ANU Press.
Duffy, R., Massé, F., Smidt, E., Marijnen, E., Büscher, B., Verweijen, J., Ramutsindela, M., Simlai, T., Joanny, L., & Lunstrum, E. (2019). Why we must question the

militarisation of conservation. *Biological Conservation, 232*, 66–73. https://doi.org/10.1016/j.biocon.2019.01.013
Dyke, F. V., & Lamb, R. L. (2020). *Conservation Biology: Foundations, Concepts, Applications* (3rd ed., 2020 edition). Springer.
Escobar-Pemberthy, N., & Ivanova, M. (2020). Implementation of multilateral environmental agreements: Rationale and design of the environmental conventions index. *Sustainability, 12*(17), Article 17. https://doi.org/10.3390/su12177098
Felbab-Brown, V. (2017). *The Extinction Market: Wildlife Trafficking and How to Counter It*. Oxford University Press.
Folke, C. (2007). Social-ecological systems and adaptive governance of the commons. *Ecological Research, 22*, 14–15. https://doi.org/10.1007/s11284-006-0074-0
Folke, C., Biggs, R., Norström, A., Reyers, B., & Rockström, J. (2016). Social-ecological resilience and biosphere-based sustainability science. *Ecology and Society, 21*(3). https://doi.org/10.5751/ES-08748-210341
Freese, C. H., Aune, K. E., Boyd, D. P., Derr, J. N., Forrest, S. C., Cormack Gates, C., Gogan, P. J. P., Grassel, S. M., Halbert, N. D., Kunkel, K., & Redford, K. H. (2007). Second chance for the plains bison. *Biological Conservation, 136*(2), 175–184. https://doi.org/10.1016/j.biocon.2006.11.019
Geist, V. (1994). Wildlife conservation as wealth. *Nature, 368*(6471), 491–492. https://doi.org/10.1038/368491a0
Guala, F. (2016). *Understanding Institutions: The Science and Philosophy of Living Together* (1st ed.). Princeton University Press.
Gunningham, N., & Sinclair, D. (2017). Smart regulation. In P. Drahos (Ed.), *Regulatory Theory* (pp. 133–148). ANU Press. www.jstor.org/stable/j.ctt1q1crtm.16
Harfoot, M., Glaser, S. A. M., Tittensor, D. P., Britten, G. L., McLardy, C., Malsch, K., & Burgess, N. D. (2018). Unveiling the patterns and trends in 40 years of global trade in CITES-listed wildlife. *Biological Conservation, 223*, 47–57. https://doi.org/10.1016/j.biocon.2018.04.017
Hepworth, R. (1998). The Independent review of CITES. *Journal of International Wildlife Law & Policy, 1*(3), 412–432. https://doi.org/10.1080/13880299809353911
Hiller, C., & 't Sas-Rolfes, M. (2025). Systematic review of the impact of restrictive wildlife trade measures on conservation of iconic species in southern Africa. *Conservation Biology, 39*(1), e14262. https://doi.org/10.1111/cobi.14262
Hinsley, A., & 't Sas-Rolfes, M. (2020). Wild assumptions? Questioning simplistic narratives about consumer preferences for wildlife products. *People and Nature, 2*(4), 972–979. https://doi.org/10.1002/pan3.10099
Holmes, G., Sandbrook, C., & Fisher, J. A. (2017). Understanding conservationists' perspectives on the new-conservation debate. *Conservation Biology, 31*(2), 353–363. https://doi.org/10.1111/cobi.12811
Hutton, J., & Dickson, B. (2000). *Endangered Species, Threatened Convention: The Past, Present and Future of CITES, the Convention on International Trade in Endangered Species of Wild Fauna and Flora*. Earthscan.
Jepson, P., & Ladle, R. J. (2010). *Conservation: A Beginner's Guide*. Oneworld Publications.
Kahneman, D. (2013). *Thinking, Fast and Slow* (1st ed.). Farrar, Straus and Giroux.
Kareiva, P., & Marvier, M. (2012). What is conservation science? *BioScience, 62*(11), 962–969.
Kontoleon, A., Pascual, U., & Swanson, T. M. (2007). *Biodiversity Economics: Principles, Methods and Applications*. Cambridge University Press.

Kosloff, L. H., & Trexler, M. C. (1987). The convention on international trade in endangered species: Enforcement theory and practice in the United States. *Boston University International Law Journal, 5*(2), 327–362.
Kriebel, D., Tickner, J., Epstein, P., Lemons, J., Levins, R., Loechler, E. L., Quinn, M., Rudel, R., Schettler, T., & Stoto, M. (2001). The precautionary principle in environmental science. *Environmental Health Perspectives, 109*(9), 871–876. https://doi.org/10.2307/3454986
Lauth, H.-J. (2015). Formal and informal institutions. In J. Gandhi & R. Ruiz-Rufino (Eds.), *Routledge Handbook of Comparative Political Institutions* (pp. 56–69). Taylor & Francis Group. http://ebookcentral.proquest.com/lib/oxford/detail.action?docID=2011232
Lorimer, J. (2015). *Wildlife in the Anthropocene: Conservation after Nature* (Illustrated ed.). University of Minnesota Press.
Lueck, D. (2002). The extermination and conservation of the American bison. *The Journal of Legal Studies, 31*(S2), S609–S652. https://doi.org/10.1086/340410
Lueck, D., & Miceli, T. J. (2007). Chapter 3 Property law. In A. M. Polinsky & S. Shavell (Eds.), *Handbook of Law and Economics* (Vol. 1, pp. 183–257). Elsevier. https://doi.org/10.1016/S1574-0730(07)01003-1
Lueck, D., & Torrens, G. (2020). Property rights and domestication. *Journal of Institutional Economics, 16*(2), 199–215. Cambridge Core. https://doi.org/10.1017/S1744137419000390
Lyster, S. (1985). *International Wildlife Law: An Analysis of International Treaties Concerned with the Conservation of Wildlife*. Cambridge University Press.
Mace, G. M. (2014). Whose conservation? *Science, 345*(6204), 1558–1560. https://doi.org/10.1126/science.1254704
Mahoney, S. P., & Geist, V. (Eds.). (2019). *The North American Model of Wildlife Conservation*. Johns Hopkins University Press.
Marris, E. (2011). *Rambunctious Garden: Saving Nature in a Post-Wild World*. Bloomsbury Publishing USA.
Meeks, D., Morton, O., & Edwards, D. P. (2024). Wildlife farming: Balancing economic and conservation interests in the face of illegal wildlife trade. *People and Nature, 6*(2), 446–457. https://doi.org/10.1002/pan3.10588
Milner-Gulland, E. J., & Mace, R. (1998). *Conservation of Biological Resources* (1st ed.). Wiley-Blackwell.
Nash, R. F., & Miller, C. (2014). *Wilderness and the American Mind: Fifth Edition*. Yale University Press.
North, D. C. (1991). Institutions. *The Journal of Economic Perspectives, 5*(1), 97–112.
Norton, B. G. (2005). *Sustainability: A Philosophy of Adaptive Ecosystem Management* (New ed.). University of Chicago Press.
Oldfield, S. (Ed.). (2003). *The Trade in Wildlife: Regulation for Conservation*. Earthscan.
Ostrom, E. (1990). *Governing the Commons: The Evolution of Institutions for Collective Action*. Cambridge University Press.
Ostrom, E. (2010). Beyond markets and states: Polycentric governance of complex economic systems. *The American Economic Review, 100*(3), 641–672. www.jstor.org/stable/27871226
Princen, T. (1994). The ivory trade ban: NGOs and international conservation. In T. Princen & M. Finger (Eds.), *Environmental NGOs in World Politics: Linking the Local and the Global* (pp. 121–159). Routledge.
Reeve, R. (2002). *Policing International Trade in Endangered Species: The CITES Treaty and Compliance*. Earthscan.

Reyers, B., Folke, C., Moore, M.-L., Biggs, R., & Galaz, V. (2018). Social-ecological systems insights for navigating the dynamics of the Anthropocene. *Annual Review of Environment and Resources, 43*, 267–289. https://doi.org/10.1146/annurev-environ-110615-085349

Rockström, J., Kotzé, L., Milutinović, S., Biermann, F., Brovkin, V., Donges, J., Ebbesson, J., French, D., Gupta, J., Kim, R., Lenton, T., Lenzi, D., Nakicenovic, N., Neumann, B., Schuppert, F., Winkelmann, R., Bosselmann, K., Folke, C., Lucht, W., … Steffen, W. (2024). The planetary commons: A new paradigm for safeguarding Earth-regulating systems in the Anthropocene. *Proceedings of the National Academy of Sciences, 121*(5), e2301531121. https://doi.org/10.1073/pnas.2301531121

Searle, J. R. (2005). What is an institution? *Journal of Institutional Economics, 1*(1), 1–22. https://doi.org/10.1017/S1744137405000020

Simon, H. A. (1955). A behavioral model of rational choice. *The Quarterly Journal of Economics, 69*(1), 99–118. https://doi.org/10.2307/1884852

Singer, P. (1975). *Animal Liberation: A New Ethics for Our Treatment of Animals*. New York Review.

Smith, V. L., & Wilson, B. J. (2019). *Humanomics: Moral Sentiments and the Wealth of Nations for the Twenty-First Century*. Cambridge University Press.

Soulé, M. E. (1985). What is conservation biology? *BioScience, 35*(11), 727–734. https://doi.org/10.2307/1310054

Spash, C. L., & Hache, F. (2022). The Dasgupta Review deconstructed: An exposé of biodiversity economics. *Globalizations, 19*(5), 653–676. https://doi.org/10.1080/14747731.2021.1929007

Steffen, W., Crutzen, P. J., & McNeill, J. R. (2007). The Anthropocene: Are humans now overwhelming the great forces of nature. *AMBIO: A Journal of the Human Environment, 36*(8), 614–621. https://doi.org/10.1579/0044-7447(2007)36[614:TAAHNO]2.0.CO;2

Swanson, T. M. (1994). The economics of extinction revisited and revised: A generalised framework for the analysis of the problems of endangered species and biodiversity losses. *Oxford Economic Papers, 46*, 800–821.

't Sas-Rolfes, M., & Challender, D. W. S. (2020). Evaluating the impact of pangolin farming on conservation. In D. W. S. Challender, H. C. Nash, & C. Waterman (Eds.), *Pangolins: Science, Society and Conservation* (pp. 517–527). Academic Press.

't Sas-Rolfes, M., Challender, D. W. S., Hinsley, A., Veríssimo, D., & Milner-Gulland, E. J. (2019). Illegal wildlife trade: Scale, processes, and governance. *Annual Review of Environment and Resources, 44*(1), 201–228. https://doi.org/10.1146/annurev-environ-101718-033253

't Sas-Rolfes, M., Challender, D. W. S., & Wainwright, L. (2025). Playing the CITES game: Lessons on global conservation governance from African megafauna. *Environmental Policy and Governance, 35*(1), 48–63. https://doi.org/10.1002/eet.2123

't Sas-Rolfes, M., & Emslie, R. (2024). African rhino conservation and the interacting influences of property, prices, and policy. *Ecological Economics, 220*, 108123. https://doi.org/10.1016/j.ecolecon.2024.108123

't Sas-Rolfes, M., Emslie, R., Adcock, K., & Knight, M. (2022). Legal hunting for conservation of highly threatened species: The case of African rhinos. *Conservation Letters, 15*(3), e12877. https://doi.org/10.1111/conl.12877

Tallis, H., & Lubchenco, J. (2014). Working together: A call for inclusive conservation. *Nature, 515*(7525), 27–28. https://doi.org/10.1038/515027a

Taylor, W. A., Lindsey, P. A., Nicholson, S. K., Relton, C., & Davies-Mostert, H. T. (2020). Jobs, game meat and profits: The benefits of wildlife ranching on marginal lands in South Africa. *Biological Conservation, 245*, 108561. https://doi.org/10.1016/j.biocon.2020.108561

Thaler, R. H. (2016). *Misbehaving: The Making of Behavioral Economics* (Reprint ed.). W. W. Norton & Company.

Trouwborst, A., Loveridge, A. J., & Macdonald, D. W. (2020). Spotty data: Managing international leopard (Panthera pardus) trophy hunting quotas amidst uncertainty. *Journal of Environmental Law, 32*(2), 253–278. https://doi.org/10.1093/jel/eqz032

van Kooten, G. C., & Bulte, E. H. (2000). *The Economics of Nature: Managing Biological Assets*. Wiley.

Weddell, B. J. (2023). *Conservation in the Context of a Changing World: Concepts, Strategies, and Evidence* (1st ed.). Cambridge University Press.

Wijnstekers, W. (2018). *The Evolution of CITES: A Reference to the Convention on International Trade in Endangered Species of Wild Fauna and Flora* (11th ed.). International Council for Game and Wildlife Conservation. https://cites.org/sites/default/files/common/resources/The_Evolution_of_CITES_2018.pdf

Wilson, E. O. (Ed.). (1988). *Biodiversity*. National Academies Press. www.nap.edu/catalog.php?record_id=989

Wyatt, T. (2021). *Is CITES Protecting Wildlife?: Assessing Implementation and Compliance* (1st ed.). Routledge.

3 Markets for Wildlife Goods

Markets can vary from simple, irregular, and informal trading events (transactions) between two parties to complex and coordinated formal mechanisms. They are shaped by the interactions of people, responding to various institutional and other incentives. All markets depend on three institutional elements: property, trading rules, and prices. How these elements emerge and function will depend on contextual factors, including geography, history, and culture. However, all markets share some essential features that are analytically useful. To lay the foundations of our economic framework, we consider the basic elements of a simple wildlife trading relationship and how these might change as the market evolves and expands into a trade chain. We also consider the varying roles that market participants may play in this trade chain and relate these to the concepts of supply and demand that underpin our analytical approach. As we do this, we consider various real-world examples of wildlife markets. We start by describing some of these.

Wildlife Markets

The wildlife trade consists of a combination of legal and illegal markets. The distinction is not always clear because many legal market participants disregard regulations (especially those of relatively minor consequence) and frequently do so unintentionally[1] (through ignorance or error). The magnitude of the wildlife trade is difficult to measure, whether expressed in terms of organisms extracted from the wild or monetary value. To some extent, CITES trade data provides some clues. However, it only effectively captures Appendix II listed species crossing international boundaries legally and misses domestic trade, non-listed species, and illegal trade[2] (which is concealed). There are also other limitations related to how CITES transactions are recorded and can thus be interpreted (Challender et al., 2022). The USA also maintains a wildlife trade database (known as LEMIS[3]), which provides some indication of the scale of legal trade but relatively little information on its conservation impact (Marshall et al., 2025). The extent of both legal and illegal wildlife markets is substantial,[4] but information remains inadequate, eliciting concern from

DOI: 10.4324/9781003286318-3

conservationists. Attempts at placing a monetary value on the trade, especially the illegal component, are fraught with difficulty and lack credibility ('t Sas-Rolfes et al., 2019).

The largest volumes of wildlife trade by far relate to timber and fisheries products (Broad et al., 2003). Despite this, and possibly because these trades are managed mainly by sizable legally registered businesses and involve less visibly charismatic species, they do not attract as much public attention as many other wild species trades. Conservation researchers are increasingly becoming aware of a broader range of previously overlooked but affected species (Hinsley et al., 2024), and we provide some categorical examples below. A select group of highly charismatic large mammals continues to dominate wildlife trade policy agendas, and we prioritize providing examples of these.

Examples

Perhaps the most iconic form of wildlife trade is that of elephant ivory,[5] with an elephant tusk forming part of the CITES logo. The ivory trade dates back far in history but likely peaked during the nineteenth century when European colonists brought guns to Africa, driving the large-scale slaughter of elephants. At the inception of CITES, African elephants were placed on Appendix II, but this did not initially stem poaching to the satisfaction of influential conservationists, who successfully lobbied for an up-listing to Appendix I in 1989 to establish an international ivory trade ban. Since then, ivory trade policy has been somewhat complicated and contested, and we discuss this further in Chapter 4.

Another prominent trade is that of rhino horn (already discussed in Chapter 1), which has persisted illegally despite an international ban that has been in place since 1977. Both rhino horn and ivory continue to command sufficiently high prices on illegal markets to incentivize continued poaching and trafficking activity. Pangolins have recently risen to prominence as "the world's most trafficked mammal," with international trade banned since a 2016 up-listing of all pangolin species to CITES Appendix I. They are sought for their meat and their scales, which are used as a traditional Chinese medicine ingredient. Also prominent is the controversial trade in primates for pets, entertainment, laboratory experiments, food, and some traditional medicine.

Large terrestrial carnivores such as big cats and bears tend to capture the public's imagination and thus feature visibly in trade. They are sought as live specimens (for exotic pets, zoos, circuses, and breeding) and for their body parts. Skins, claws, and teeth are sought for ornamental purposes, and big cat bones and bear bile (gallbladders) are sought for traditional medicinal purposes in Asia. Although less prominent, many smaller terrestrial carnivores are also traded, frequently as pets and for their fur. Increasingly, participants in the carnivore trade have turned to commercial captive breeding to supply what are otherwise comparatively scarce animals in the wild. Terrestrial herbivores are also harvested for trade, typically for food (meat), hides, horns, and wool; increasingly, some are also farmed or otherwise ranched in free-ranging conditions.

Beyond mammals, there are substantial trades in live birds, which are widely kept as pets—notable examples include the Latin American parrot trade and the Indonesian songbird trade. The bird trade also includes meat, eggs, leathers, and feathers, with the ostrich trade being a prominent example, now provided mainly by farmed birds. There are also niche markets in exotic products such as hornbill ivory and a large-scale trade in bird eggshells for ornamental and collection purposes. Much of the bird trade is legal, but there are also sizeable illegal elements. The same applies to a substantial trade in reptiles and amphibians, which are variously sought as pets and for their skins, shells, and flesh. Crocodilian and snake skins are especially sought as high-value fashion materials for shoes, bags, jackets, and belts, and their trade is managed by large corporations in the luxury goods sector.

Larger marine species, such as cetaceans, sharks, seals, and turtles, are sought for their meat and body parts, such as shark fins, which are considered a delicacy in some Asian markets. Cetaceans and seals are also captured and trained for entertainment. A plethora of smaller marine species, including fish, mollusks, crustaceans, seahorses, and corals, are variously traded as pets, food, medicines, and ornaments, among other things. Some marine and aquatic species are linked to problematic high-value illegal markets, also involving controversial attempts to farm them—examples include abalone, totoaba, European eels, and sturgeon. On land, smaller animals such as arachnids and insects (notably butterflies) are also widely traded for personal collections, food, and other products, and many of them are also farmed.

Finally, while less likely to attract attention and stir public emotion, there is a substantial trade in wild plants and fungi, including significant illegal activity (Margulies et al., 2019). The most obvious plant product is timber, with many tree species affected; rosewoods, cedar, and agarwood are of the greatest concern as the highest-value illegal trades, dominated by organized crime. Orchids constitute a highly significant plant group, being traded for ornamental, food, and medicinal reasons, and account for the largest proportion of CITES-listed species (Hinsley et al., 2018). Diverse plants are harvested for all their parts, generating products such as roots, bulbs, stems, wood, leaves, flowers, fruits, nuts, seeds, starches, sugars, oils, gums, resins, waxes, gels, pigments, and fibers. These are used for construction, fuel, clothing, decoration, food, medicine, cosmetics, and other products, such as perfumes.

Live plants may be taken from the wild and captive bred, with cycads providing examples of species harvested to extinction in the wild (in South Africa) and now only surviving in captivity. Cacti and other succulent species are heavily targeted by plant poachers, and there has been a recent proliferation of organized illegal activity in South Africa around this trade (Hübschle & Margulies, 2024). Fungi are traded mainly for food and medicinal purposes.

Although under-researched, there is evidence of increasing trade volumes leading to the depletion of certain fungal species (Oyanedel et al., 2024).

Wildlife Market Evolution

Even the most basic wildlife product market, if reduced to a single transaction (a "one-off trade"), involves a buyer and a seller. Assuming the buyer is the end user of the wildlife product, we can refer to them as the "consumer" and represent their side of the transaction as the "demand" side. Assuming that the seller of the product is the same person who harvested it from the wild, we represent their side as the "supply" side. By the standard commercial definition, trade takes place when the seller transfers ownership of the product to the buyer, typically receiving some kind of payment in return. The value of that payment (as measured in monetary terms) constitutes the "market price" of the transaction. Note that this price does not amount to an objective universal value[6] of the product in question; rather, it simply represents the subjective value that both the buyer and seller place on obtaining or selling that specific product at the time of that specific transaction. Stated another way, the market price is the agreed value that both parties to the transaction place on the supply meeting the demand. From this basic building block of a single transaction, we can develop models of how markets evolve and assume certain structures.

From the initial transaction, a wildlife market can grow and evolve in several ways. For example, (1) the two parties can agree to trade another or more items of the same product; (2) more parties can agree to engage in buying or selling items of the same product; or (3) the product buyer, instead of being the end user of the product, could act as an intermediary, and sell the product onward to a different end user, thereby forming a nascent "trade chain." In each of these cases, we see the market expanding with additional transactions, which may or may not take place at the same market price as the initial transaction.[7] The unfolding of such growth and evolution through repeated transactions will yield multiple potential price points over time in what we call the market discovery process. However, over time, we would also likely see the emergence of patterns that would yield average or standardized market prices for essentially similar products traded under similar circumstances.

A key determinant of how such wildlife markets might evolve would be the underpinning institutional arrangements that enable the possession and transfer of the product(s) in question. A first factor would be the extent to which the product would be legally and socially recognized as property—in other words, the strength of the property right, as discussed in Chapter 2. Stronger property rights increase the probability of trade—legal recognition of property provides assurance to buyers and sellers that their transaction contract is officially enforceable. However, trade can still occur when property rights are weak—and even when legal ownership is completely absent, as we see with illegal trade. When property rights are poorly enforced by governments, possession may be all that matters to those doing the trading. A second institutional factor would concern trading rules and regulations set by governments—i.e., whether these outlaw, discourage, or enable trading activity. A third institutional factor would be the extent to which a mutually agreeable and appealing form of money exists—markets are more likely to develop when there is a shared (ideally widely recognized) and stable currency that both buyers and sellers can access and easily use (e.g., US dollars).

Another key determinant of market evolution is the role of entrepreneurs. As noted in the previous chapter, entrepreneurs seek out and pursue economic opportunities. This may involve trying to strengthen property rights—for example, by claiming an open-access good or by enabling the laundering of illegal (stolen) goods to become reclassified as legal goods that are more easily sold. Most entrepreneurs will also seek to reduce transaction costs—in other words, they will try to find ways to do things that are cheaper and more efficient. This may involve lobbying for regulatory change, but it may also involve innovations such as better use of technology to improve efficiency or finding easier trade routes. Entrepreneurs also often aim to increase their profit margins by seeking out potential new product consumers and advertising their products to them (the activity known as "marketing") and by adding enhancements to their products, either by modifying and improving the products themselves or by using creative "branding" to differentiate the products from others or make them otherwise more socially desirable.

Entrepreneurs constantly seek "arbitrage" opportunities, which exist when there is a significant difference in the price of the same good between two markets, which they can exploit for their own benefit. Under these circumstances, they may become intermediary traders, buying from the seller with the lower price and then selling to the market with the higher price, extracting a profit for themselves from the so-called "mark-up" (being the difference between the two prices). Another form of arbitrage exists when there are varying expectations of future prices. In the case of storable goods, entrepreneurs may speculate that prices will rise faster than average price increases for other goods[8] and thus buy up stocks of the product to store and sell for a profit at a future stage. We discuss this further in Chapter 7.

Markets thus evolve out of a combination of interacting institutional arrangements and entrepreneurial activity. Entrepreneurs may emerge from anywhere—they may be original suppliers, consumers, or third parties who see opportunities, including politicians. Entrepreneurs are reasonably abundant in society and have driven economic development throughout history. Their actions are the reason that simple voluntary exchanges of harvested wild products between neighboring primitive human communities eventually evolved into the complex set of arrangements that characterize today's global wildlife trade. Human society has also created institutions such as money, legal systems, and governments to facilitate repetitive transactions by embedding expectations of how they would turn out for both parties. This has resulted in the structural characteristics of modern wildlife markets, which we now discuss.

Structures and Processes

In rural frontier areas, where people live near wildlife, we may still find simple direct markets whereby one person harvests a wild species and sells it directly to another to consume. However, in almost all cases, harvested wildlife specimens will change hands multiple times before reaching the ultimate consumer, likely being stored or modified along the trade chain. Extending our simple model of a market with only a harvester/seller and consumer/buyer, we can insert a third party, the intermediary,[9] to derive a basic trade chain model. In that case, there are two transaction points, from harvester to intermediary and from intermediary to consumer. At each transaction point, the seller represents the supply, and the buyer represents the demand. We can insert as many intermediaries as we like into this trade chain,[10] and the principle will remain the same for each successive transaction. There will always be a supply (seller) side and a demand (buyer) side, and each will have the same motivation: to achieve the best possible price for themselves. For sellers, it will be the highest possible price, and for buyers, it will be the lowest. The final trading price will be achieved by negotiation.[11]

In the next chapter, we will explore how supply and demand are determined and how they interact in more detail. In our framework, the final trading price will emerge from the "equilibrium" point between supply and demand. Skeptical ecologists and other systems thinkers might argue that actual stable equilibrium points rarely exist in the chaotic natural world; however, empirical research confirms that, much like gravity, the tendency of markets to converge toward equilibrium (often described as "market forces") is powerful.[12] Market forces are typically only thwarted by other forces that specifically prevent competition or otherwise intervene in markets, usually with the backing of governments (with their socially sanctioned monopoly on the ultimate use of physical force, including violence, to constrain human behavior).

Two factors influence these market structures and processes, both as consequences of human nature. The first is human preferences for stability and predictability (being the primary motivation for creating institutions). For example, humans have an aversion to constantly changing prices, and this can cause market prices to become somewhat "sticky." The second, as Adam Smith already observed, is that people are also averse to experiencing intense open competition and will attempt to gain and maintain privileged monopoly power whenever possible. They may achieve this by obtaining exclusive property rights over physical goods and even ideas (as "intellectual" property, secured through patents) or by securing other barriers to entry into markets (by, e.g., lobbying governments to grant exclusive licenses or establish onerous regulations and standards).

How do the structural factors and processes we have described manifest in real-world examples? Let us consider four different wildlife product market categories[13]: (1) Living organisms, (2) Food, (3) Traditional medicines, and (4) Fashion items and ornaments.

Living organisms are sought as pets, entertainment, and for ornamental purposes. Many live specimens are challenging to harvest and transport—this typically requires specialist skills. Handling large animals, especially, requires professional veterinary and logistical support. Consequently, large animal supply is dominated by a relatively small number of operators, trade chains tend to be short, and this trade is, in theory, easily controlled and monitored. In practice, large animal buyers tend to be powerful elites (e.g., zoos and wealthy collectors) who can afford to influence regulatory control or pay bribes to high-ranking government officials to facilitate certain transactions.

For smaller animals and plants, species vary in how easily they can be transported or concealed, which determines whether trade can take place more sporadically and less formally. There are specialist traders in certain types of species, and others may be traded by amateurs, but to ensure that specimens are kept alive, trade chains typically need to be short and fast-moving unless there is potential to maintain specimens in captive facilities. Captive facilities are game changers, especially if they enable captive breeding, in which case other issues come into play. As with large animal handling, captive facilities ideally require some skilled management to provide adequate sustenance, maintain welfare standards, and prevent disease, all of which are costly. In practice, there are many amateur (and sometimes illegal) captive facilities with poor standards, which is an issue of great concern to the animal protection movement.

Food markets vary from the trade of harvested animal flesh to harvested plant material. Many food items are perishable and have limited shelf lives—as with live animals, this places limitations on trade chains. While harvesting may be spread across many individuals or small groups, collection and transport to retailers are typically more concentrated. Freshly harvested wild foods either need to reach consumers quickly or must be refrigerated, dried, or otherwise processed. In some cases, this is easily done by amateurs, but to scale up such markets typically requires more technical and logistical skills and is more likely to become professionalized and concentrated around intermediaries, with the most obvious example of this being the large companies that manage ocean harvests for international trade. Markets for wild meat tend to be less organized and dispersed, and there is much illegal activity, but this sector is growing and becoming increasingly professionalized as the global demand for wild meat is rapidly growing (with more affluent consumer preferences shifting somewhat away from factory-farmed meat of domestic species).

Traditional medicine markets vary from local and informal to China's highly organized formal Traditional Chinese Medicine (TCM) production and distribution system. Market structure varies greatly with species, but for many widely consumed products (mainly plants), large TCM corporations have vertically integrated by creating captive production facilities (i.e., wildlife farming). Official TCM distribution in China takes place through a state clinic system, which sets selling prices. Prestigious animal products in TCM include rhino horn and tiger bone, both of which are no longer legally available since China agreed to close its legal domestic markets in 1993. Other animal products, such as saiga horn, pangolin scales, and bear bile, remain legal in some domestic Asian markets; this is controversial and subject to ongoing negotiations either because wild populations of these species are subject to continued high poaching levels (especially pangolins) or because they are farmed, eliciting welfare concerns (e.g., bears). Informal and illegal TCM markets drive demand for endangered animal products, such as rhino horn, pangolin scales, saiga horn, and big cat bones, which are provided through various illegal supply channels. These include trafficked animals poached in Africa and, in the case of big cat bones, from captive facilities. At least some of these products (e.g., rhino horn and pangolin scales) are transported internationally by organized crime syndicates acting as specialized intermediaries, which may effectively enjoy some (unofficial) monopoly power.

Durable wildlife products (e.g., wood, bones, horns, hides, and teeth) provide materials for a range of utility and decorative goods and are especially popular in fashion (clothing and accessories) and as ornaments (e.g., carvings). Fashion goods and ornaments typically have an added element of human crafting, which can significantly boost their final market value. Markets for such goods tend to be more personalized with a prestige element[14] but also potentially somewhat fickle and more easily influenced. In many instances, these goods signal social status and, in the case of rare and endangered species, may signal immunity from regular law enforcement or a form of cultural affiliation (perhaps even in defiance of other cultures). Some durable wildlife products may also serve as speculative investments, a potential store of value, and in barter trade as currency alternatives to money.

As a classic example, the market for elephant ivory illustrates some other issues relevant to this category of wildlife product market. One is the potential for long-term competition with holders of older stocks, with some well-preserved ivory lasting for centuries. Another is the issue of competition from—and potential replacement with—substitute and fake goods (e.g., ivory from other species such as hippos, ivory from extinct mammoths, plastics, and other synthetics). For other species, this may include farmed goods (e.g., crocodile skins and furs). Depending on context, there may be incentives to distinguish such goods from others or disguise them as others. When such markets are heavily regulated, as with ivory, this can create opportunities to create monopolistic intermediary structures, which may or may not benefit conservation, depending on how they are positioned. Monopolistic intermediaries were established for two "one-off" international ivory sales under CITES (discussed further in Chapter 4), and other international markets are similarly structured, such as that for vicuña wool, which is controlled by a single international company. Although providing seemingly beneficial control through a single market channel, a potential disadvantage of such structures is that the intermediaries may claim most of the profits, with comparatively little going to the original suppliers, whether wildlife managers or local communities.

Understanding the above elements helps us make sense of the wildlife markets we observe. Not all existing wildlife markets are efficient, fair, or sustainable. They are products of past entrepreneurial actions within institutional constraints and shaped by circumstantial factors. Future events causing circumstantial change and entrepreneurial actions will further modify these markets. Conservationists may want to consider some of the issues with existing wildlife markets and how we can leverage an economic understanding of market structures and processes to move toward systems that achieve a preferred balance of conservation and economic goals. A good way to start this process is to think about the circumstances and incentives facing each of the different categories of market actors.

Market Actors

Market actors account for the factors that underlie supply and demand in every transaction. Based on their position in the trade chain and the associated costs and benefits of trading in that position, market actors face certain incentives. Understanding their position and these incentives is the key to understanding what drives wildlife trade and how we might want to engage with it for the benefit of conservation (or other social goals). As previously discussed, we can divide market actors into three broad categories: the initial wildlife harvesters, product end users ("consumers"), and any intermediaries. However, the supply side is slightly more complicated. For the purposes of our analysis, focused on preventing poaching (for the sake of conservation), we can identify three relevant supply-side sub-categories: legal producers, poachers, and other illegal producers. We discuss the relevance of these differences below, but first, we consider consumers.

Consumers

As the end users of all wildlife products, consumers ultimately drive the wildlife trade. Without consumer demand, there would be no reason to trade.[15] In economic terms, consumer demand comprises two elements: the consumers' preferences to buy the product and their available resources (e.g., money). Consumer preferences are based on their motivations, which can be complex and variable (Veríssimo et al., 2020). As indicated in our earlier examples of species in trade and product categories, consumers may desire wildlife products to meet basic needs (e.g., food, medicine, and clothing) or for reasons that have more to do with factors such as pleasure, cultural tradition, or social status, and often a combination of these. Consumer motives may be difficult to untangle.

Wildlife trade researchers have identified five broad motivational categories: experiential, social, functional, financial, and spiritual—and many further sub-categories (Thomas-Walters et al., 2021). However, for our purposes, the exact nature and relationship of the complex motivational factors underlying preferences are not that important. What does matter is whether the preferences overall are stable and persistent or fickle—this will depend on various things, including whether buying the product becomes illegal[16] (and therefore risky) or socially unacceptable. The sum of factors leading to a preference can be thought of as the total benefit the consumer will expect to gain from a purchase. We would then expect them to proceed with a purchase if the product is available and they can afford it (i.e., the benefits exceed the costs). This will then determine the maximum price they are willing to pay for it. One final consideration is the extent to which consumers have other options to satisfy their relevant desires. This, combined with the amount of money they can afford to spend, will determine their sensitivity to changes in the market price (i.e., demand elasticity).

Legal Producers

There are several different potential sub-categories of legal wildlife product providers. Traditionally, these would have been hunters and harvesters who were legally entitled to extract organisms from the wild, often in the context of open-access property rights. This is still the case for marine and coastal species in situations where local people may have harvesting rights. However, since true open access conditions are increasingly rare, especially on land, we need to examine the legal ownership of wildlife to determine the identity of the original legal producer. In many countries, governments and their agencies are the designated legal wildlife custodians, and they may sometimes decide to harvest and sell wild-extracted products from state-protected areas. As an example, in several African countries, government agencies have in the past sold their stocks of accumulated products such as elephant ivory and rhino horn, as well as meat and hides harvested from various animals that were culled for reasons of population control for ecosystem management. Many agencies also sell wild plants, for example, through state-run nurseries.

Private and communal landowners are another source of production and, depending on the extent to which this is allowed by government agencies, may harvest and sell products from their land. In South Africa, for example, there is extensive private ownership of wildlife land, managed across a spectrum of conservation to commercial objectives, and private ownership and sale of many wild species is permitted. Consequently, there is a vibrant trade in both live and harvested specimens. However, in recent years, there has also been a trend (echoed worldwide) for wild species to be more intensively managed in conditions that increasingly resemble commercial ranching or farming rather than conservation. Wildlife farmers and commercial captive breeders thus represent another form of legal producer. As a final emerging category, some wildlife products may be produced through genetic modification or synthetic biology to provide an alternative form of legal supply. The extent of this is currently limited but will likely grow in the future.

Different sub-categories of legal producers will face different costs, which will also affect the selling price they would be willing to accept (to cover those costs). Traditional legal harvesters in open access conditions will bear the costs of equipment, supplies, and time needed to harvest, but typically little else. By contrast, government agencies and private landowners must typically bear management costs to maintain ecosystem integrity (habitat quality) and security to deter illegal harvesting. This would include the costs of harvesting (including culling). As management increases in intensity and tends toward farming, additional costs will surface, such as fencing, supplementary nutrition and water, disease management, and catering to other welfare needs of the species in question. Synthetic products will obviously incur a different set of costs related to research and manufacture. Hence, the supply costs of legal producers might vary significantly between those involved in traditional wild-harvesting, state production, private production, farming, and manufacturing. As a final important note, legal producers with ownership rights over free-ranging wildlife can reinvest the proceeds from product sales directly into further "production," thereby benefiting conservation—this is in contrast with the other sub-categories of traditional open-access harvesters, farmers, and manufacturers, who will only do so indirectly if explicitly taxed for that purpose.

Poachers

Poachers can be defined simply as illegal harvesters. They range from local people whose traditional harvesting practices have been outlawed to professionals who target high-value species in the open seas or on public or private land. Many poachers fall into the first category (i.e., "subsistence" poaching), being rural people skilled at traditional hunting or harvesting. For such people, the benefits of poaching may include meeting personal nutritional, medicinal, cultural, or status needs, as well as earning profits from selling the poached item. In some instances (in the case of larger dangerous animals), rural poachers may be motivated mainly by retaliation following a human–wildlife conflict incident (e.g., loss of crops, livestock, or a family member), with a subsequent sale of harvested product being incidental. As with other traditional harvesters, their costs comprise equipment, supplies, and time, to which they must add an expected cost of incurring a penalty for breaking the law. This expected cost (being the combined perceived probability of detection, apprehension, prosecution, and punishment multiplied by the full economic cost of incurring that punishment) must be added to cost assessments for all illegal actors.[17]

For more valuable species, poaching may evolve from subsistence-motivated and incidental to commercially motivated and targeted. If the rewards are high enough, poachers may become organized into gangs, which tends to happen for high-value and high-stakes species. The composition of such gangs will vary with context, with different role-players within the gang potentially receiving different levels of compensation. Gangs can include desperately poor and relatively unskilled members as well as less desperate and more skilled members, who would receive a larger portion of the reward. For some species, gangs might use sophisticated and expensive technologies (e.g., automatic weapons, night-vision equipment, high-spec boats, and aircraft)—these will obviously incur greater costs, justified by the expected rewards. When high-value species are involved, such gangs may also be highly mobile, moving across vast territories and jurisdictional boundaries, and enjoy protection and backup legal assistance (if arrested) from intermediaries.

Other Illegal Producers

Other illegal producers may include wildlife owners and managers who harvest species under their control but for which product sales are not permitted. They may also include legal wildlife breeders and farmers who ignore or violate regulations in various ways, as well as captive breeding operations that are entirely illegal. Depending on their circumstances, such illegal producers will face varying costs, including the expected costs of incurring penalties. Such producers often elicit great concern from conservationists, many of whom believe that such illegal supply activity stimulates consumer demand and somehow raises the threat to wild populations of the species concerned. However, this is questionable and depends on various factors related to how such activity impacts the overall supply and demand dynamic for a given wildlife product. Since illegal producers in this category are not directly extracting the species from the wild, it is conceivable that in some cases their actions may, in fact, benefit conservation, even if not intentionally so.

Intermediaries

Intermediaries may comprise various sub-categories, with their activities potentially including trading, transporting, consolidating, storing, processing, packaging, smuggling, facilitating (e.g., paying bribes), exporting, importing, wholesaling, retailing, marketing, and investing. Their activities may cross legal boundaries (i.e., between legal and illegal) at various stages of the trade chain and are influenced by cost, benefit, and risk assessments. Intermediaries tend to be entrepreneurial profit-seekers and thus constantly search for new opportunities to reduce costs and risks and increase benefits. The ingenuity of intermediaries presents a challenge to the enforcement of wildlife trade regulation when these actors do not share the same ethical perspective as those setting the regulations. Intermediaries are frequently better informed, skilled, and experienced—and more flexible in their operations—than those tasked with enforcing the laws.

Law-breaking intermediaries can range from individual smugglers acting alone to professional international crime syndicates that specialize in the trade and transport of a wide range of illegal goods (including drugs and weapons) and human trafficking. Such syndicates often work as networks rather than hierarchies, with a diversified internal economy that includes specialized roles such as protection and (informal) enforcement, separate from the actual handling of goods (to frustrate legal enforcement efforts). They also frequently encompass corrupt government officials who facilitate their activities and assist in thwarting potential competitors. Since all these factors will vary between species and jurisdictions, it is important to consider the specific elements that make up an intermediary's cost–benefit determination for every type of transaction where they may act as either a buyer or a seller.

Intermediaries may also significantly modulate the flow from the original supply source to the end user by storing and stockpiling goods, in some cases for extended periods. This activity is typically based on speculative attempts to increase profits through arbitrage across time periods. Intermediaries who accumulate large stockpiles may also gain market (monopolistic) power. Captive facilities for live wildlife specimens can also serve this purpose. Accumulating large stocks may be risky for intermediaries dealing with illegal products, so such activity is only likely to develop in contexts where enforcement is limited or weak. When assessing the incentives to stockpile, we must also factor in the costs and risks of doing so. In some cases (at least in theory), there may be a benefit in building up a stockpile while speculating that the species in question will become extinct, even going so far as to facilitate overharvesting to gain significant monopoly power.[18] For this to happen effectively, there would need to be some vertical integration (or at least cooperation) between the stockpiling intermediary and poachers. This, and various other issues, can be considered by applying our analytical framework, which we now turn to present in the next chapter.

Notes

1 As an example, many product consumers do not know or even care much about the origins of the products they buy. We explore the implications of incomplete information further in Chapter 6 of this book.
2 Data from seizures provide us with some indication of illegal trade, but such data tend to be sporadic and subject to many biases and cannot be relied upon to measure scale.
3 "LEMIS" stands for Law Enforcement Management Information System.
4 See Hughes (2021) for some estimates of the extent of wildlife markets.
5 See Walker (2009) and Somerville (2017) for interesting, detailed accounts of this trade.

6 There is in fact no way to determine an objective universal value of any product—all value is contextual.
7 There are many further ways in which wildlife product markets can evolve—for example, by splitting existing products into more than one type, harvesting additional products from the same species, blending products from different species, or otherwise modifying products to differentiate them from competitors.
8 This can be determined by comparison with the general rate of inflation and by a measure of the opportunity cost of simply saving and holding money, known as the "interest rate."
9 We prefer this term to the alternative term "middleman."
10 For a graphic illustration of a simplified wildlife trade chain, see Figure 3 in 't Sas-Rolfes et al. (2019: 206).
11 In well-established high-volume retail markets with frequent transactions (e.g., supermarkets and some online trading platforms), there is limited scope for actual price negotiation, for practical economic reasons.
12 Equilibrium points themselves also shift over time as market circumstances change.
13 These categories have been inspired by the literature (e.g., Jiao & Lee, 2022; Mozer & Prost, 2023) but modified.
14 Hunting trophies used in taxidermy also fall broadly into this category.
15 The conservation literature discusses whether wildlife trade is "demand-driven" or "supply-driven" (e.g., McNamara et al., 2016). However, both components are necessary for trade to take place, and the role of entrepreneurs in linking supply and demand is also critical.
16 As a reminder, depending on the context, there are many instances in which consumers may be ignorant of or not care about the legal status of a wildlife product.
17 Note that, in many cases, if the perceived probability of being ultimately punished is sufficiently low, even a severe potential penalty (including death) may be heavily discounted by the person contemplating the crime.
18 This has been referred to as "banking on extinction" (Mason et al., 2012), which we discuss further in Chapter 7.

References

Broad, S., Roe, D., & Mulliken, T. (2003). The nature and extent of legal and illegal trade in wildlife. In S. Oldfield (Ed.), *The Trade in Wildlife: Regulation for Conservation* (illustrated ed., pp. 3–22). Earthscan.

Challender, D. W. S., Brockington, D., Hinsley, A., Hoffmann, M., Kolby, J. E., Massé, F., Natusch, D. J. D., Oldfield, T. E. E., Outhwaite, W., 't Sas-Rolfes, M., & Milner-Gulland, E. j. (2022). Mischaracterizing wildlife trade and its impacts may mislead policy processes. *Conservation Letters, 15*(1), e12832. https://doi.org/10.1111/conl.12832

Hinsley, A., de Boer, H. J., Fay, M. F., Gale, S. W., Gardiner, L. M., Gunasekara, R. S., Kumar, P., Masters, S., Metusala, D., Roberts, D. L., Veldman, S., Wong, S., & Phelps, J. (2018). A review of the trade in orchids and its implications for conservation. *Botanical Journal of the Linnean Society, 186*(4), 435–455. https://doi.org/10.1093/botlinnean/box083

Hinsley, A., Hughes, A., & Margulies, J. (2024). Creating a more inclusive approach to wildlife trade management. *Conservation Biology, 38*(5), e14360. https://doi.org/10.1111/cobi.14360

Hübschle, A., & Margulies, J. (2024). The need for a socioecological harm reduction approach to reduce illegal wildlife trade. *Conservation Biology, 38*(5), e14335. https://doi.org/10.1111/cobi.14335

Hughes, A. C. (2021). Wildlife trade. *Current Biology, 31*(19), R1218–R1224. https://doi.org/10.1016/j.cub.2021.08.056

Jiao, Y., & Lee, T. M. (undefined/ed). The global magnitude and implications of legal and illegal wildlife trade in China. *Oryx, 56*(3), 404–411. https://doi.org/10.1017/S0030605320000800

Margulies, J. D., Bullough, L.-A., Hinsley, A., Ingram, D. J., Cowell, C., Goettsch, B., Klitgård, B. B., Lavorgna, A., Sinovas, P., & Phelps, J. (2019). Illegal wildlife trade and the persistence of "plant blindness". *Plants, People, Planet, 1*(3), 173–182. https://doi.org/10.1002/ppp3.10053

Marshall, B. M., Alamshah, A. L., Cardoso, P., Cassey, P., Chekunov, S., Eskew, E. A., Fukushima, C. S., García-Díaz, P., Gore, M. L., Lockwood, J. L., Rhyne, A. L., Sinclair, J. S., Strine, C. T., Stringham, O. C., Tlusty, M. F., Valdez, J. W., Watters, F., & Hughes, A. C. (2025). The magnitude of legal wildlife trade and implications for species survival. *Proceedings of the National Academy of Sciences, 122*(2), e2410774121. https://doi.org/10.1073/pnas.2410774121

Mason, C. F., Bulte, E. H., & Horan, R. D. (2012). Banking on extinction: Endangered species and speculation. *Oxford Review of Economic Policy, 28*(1), 180–192. https://doi.org/10.1093/oxrep/grs006

McNamara, J., Rowcliffe, M., Cowlishaw, G., Alexander, J. S., Ntiamoa-Baidu, Y., Brenya, A., & Milner-Gulland, E. J. (2016). Characterising wildlife trade market supply-demand dynamics. *Plos One, 11*(9), e0162972. https://doi.org/10.1371/journal.pone.0162972

Mozer, A., & Prost, S. (2023). An introduction to illegal wildlife trade and its effects on biodiversity and society. *Forensic Science International: Animals and Environments, 3*, 100064. https://doi.org/10.1016/j.fsiae.2023.100064

Oyanedel, R., Levi, M., & Furci, G. (2024). A call to include fungi in wildlife trade research and policy. *Conservation Biology, 38*(5), e14340. https://doi.org/10.1111/cobi.14340

Somerville, K. (2017). *Ivory: Power and Poaching in Africa* (1st ed.). Hurst.

Thomas-Walters, L., Hinsley, A., Bergin, D., Burgess, G., Doughty, H., Eppel, S., MacFarlane, D., Meijer, W., Lee, T. M., Phelps, J., Smith, R. J., Wan, A. K. Y., & Veríssimo, D. (2021). Motivations for the use and consumption of wildlife products. *Conservation Biology, 35*(2), 483–491. https://doi.org/10.1111/cobi.13578

't Sas-Rolfes, M., Challender, D. W. S., Hinsley, A., Veríssimo, D., & Milner-Gulland, E. J. (2019). Illegal wildlife trade: Scale, processes, and governance. *Annual Review of Environment and Resources, 44*(1), 201–228. https://doi.org/10.1146/annurev-environ-101718-033253

Veríssimo, D., 't Sas-Rolfes, M., & Glikman, J. A. (2020). Influencing consumer demand is vital for tackling the illegal wildlife trade. *People and Nature, 2*(4), 872–876. https://doi.org/10.1002/pan3.10171

Walker, J. F. (2009). *Ivory's Ghosts: The White Gold of History and the Fate of Elephants*. Grove Press/Atlantic Monthly Press.

4 An Economic Model of the Market for Wildlife Goods

In this chapter, we introduce our basic economic framework, upon which all the subsequent analyses are built. This model considers a single wildlife good, by which we mean either a live specimen or specific part of a wild animal or plant. Many species yield multiple products—for example, an antelope that has been hunted might yield meat, hides, and horns—but to start we consider only a single standardized product for the sake of simplicity. Product consumers may also view legal and illegal goods differently, and we will return to this point later in this chapter and elaborate further in the next chapter when we discuss competing substitutes, but our initial model in this chapter assumes that the entire market is either legal or illegal.

The Demand Side

We start presenting our economic framework for examining wildlife trade by considering how to model the demand side of the market. This side of the market comprises the final buyers or consumers of wildlife goods. Examples would include someone who buys and eats wild-harvested fruit, fish, or meat; someone who buys a live wild-harvested reptile like a lizard or turtle to keep as a pet; someone who buys a walking stick or furniture item made of wild-harvested wood; someone who buys clothing items made from wild-harvested products such as vicuña wool or kangaroo leather; someone who buys an ornament made from coral or elephant ivory; and someone who buys powdered rhino horn or pangolin scales for traditional healing purposes.[1]

The main element for our economic model on the demand side is the *market demand curve* for a wildlife good, which describes the relationship between the price of the good and the quantity of it that the buyers in the market want to purchase. Now, a myriad of factors can influence the decision of a buyer regarding whether to purchase a wildlife good and, if so, how much of it to buy. These factors include—but are by no means limited to—how much the consumer likes the good; the price of it; how much money the consumer has; whether there are any prohibitions (legal or cultural) against the consumption of it; how many people in the consumer's network have purchased the good;

DOI: 10.4324/9781003286318-4

what kind of marketing (positive or negative) of the product exists currently; the price of similar products; and so on. It is crucial to keep in mind that the market demand curve shows what happens to the demand for the product when the price of it changes, *keeping all other factors fixed*. Therefore, the point of the market demand curve is to isolate the effect of own-price changes on demand, assuming nothing else changes.

This does not mean, however, that the effect on demand of changes in factors other than the price of the good cannot be shown using the market demand curve. In fact, any change in demand resulting from any change in any of the factors that determine demand can be illustrated using the market demand curve. The crucial distinction is as follows: When the price of a wildlife good changes—and nothing else is altered—the impact on the aggregate demand for the good is captured by a *movement along the market demand curve*. On the other hand, if some other factor—such as the income level of the consumers—changes, this will result in a *shift of the market demand curve*. Note that a shift of the market demand curve caused by a change in some factor other than the price of the wildlife good need not be a parallel one. The distinction—between a movement along the curve and a shift of the curve—in how we utilize the market demand curve to analyze the impact of changes in the factors that determine consumer demand is illustrated in Figure 4.1.

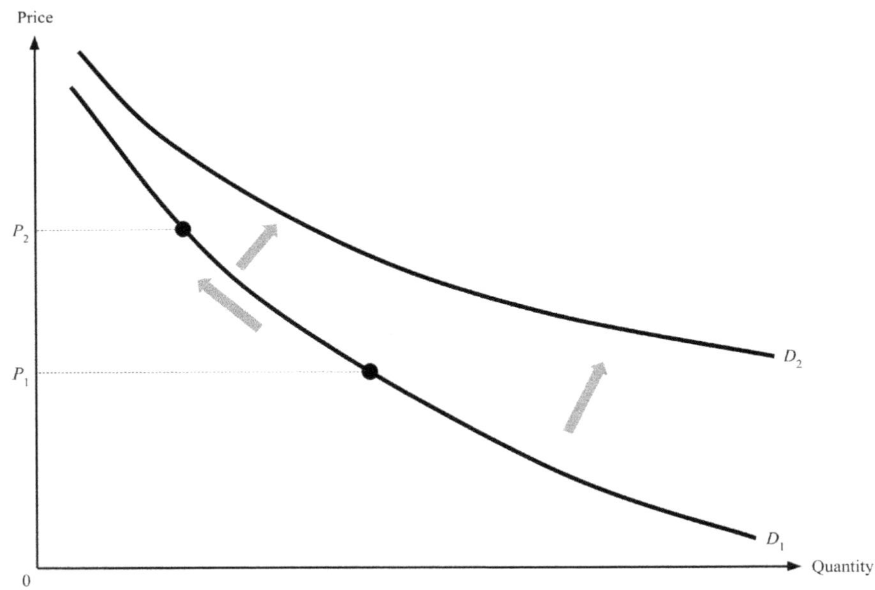

Figure 4.1 Given the market demand curve D_1, the effect on demand of an increase in the price of the wildlife good from P_1 to P_2—keeping all else fixed—is shown by the movement along curve D_1. The effect of a change in some other factor (such as income) that increases demand at every price is shown by the shift from curve D_1 to curve D_2.

The Supply Side

The sellers of wildlife goods make up the supply side of the market. Since a wildlife good can change hands several times along the supply chain before it reaches the consumer of the product, some care must be taken to distinguish between the initial producers or suppliers of the good and the intermediaries that may exist to connect these first sellers to the consumers. In this section, we will focus on the initial suppliers and discuss how to model their behavior. A discussion of the modeling of intermediaries is given in a later section.

Initial suppliers of wildlife goods may be legal or illegal. Examples of legal suppliers include official agencies responsible for managing wildlife areas, local people with harvesting rights, wildlife ranchers, and farmers of wild species. The most obvious initial illegal suppliers are poachers, who may be local people acting opportunistically, or traveling professionals who actively seek out and exploit sources for the goods. However, there are also instances of illegal cultivation or breeding operations that provide goods illegally. Again, for the purposes of our basic model, we assume only one kind of supplier (either legal or illegal).

Every supplier can be characterized by how much it costs them to provide different amounts of the wildlife good. This cost can depend on a host of factors, including how skilled the supplier is at producing or procuring the good, the capital or technology available to the supplier for production or procurement, the stock of wildlife available, and how easily accessible this stock of wildlife is. Note that policy changes such as more effective anti-poaching measures[2] can affect these costs and thereby alter suppliers' incentives to provide a wildlife good.

Much like the demand side of the market, the relevant information about the supplier's characteristics can be summarized and represented using a curve that relates the total quantity of the wildlife good supplied to the cost of supplying the next unit of the good. This is known in economics as the *marginal cost curve*. Since we would expect the most efficient producers or suppliers to enter the market and supply the good before less efficient ones who incur a higher cost to produce additional units of the good, this marginal cost curve should be upward sloping, as shown in Figure 4.2. Even if there is only a single supplier of the good in the market, we would still expect the marginal cost of production to increase with quantity. This follows since the producer—in the interest of minimizing their cost of production—would utilize the cheapest or most efficient means of producing the good at their disposal, and move on to other—i.e., more costly—means only when cheaper alternatives are no longer possible or available.

If there are no intermediaries in the market so that the initial producers sell the wildlife good directly to consumers, and if the market is perfectly competitive—i.e., each producer is small relative to the entire market and therefore has little market power[3]—then the marginal cost curve of the producers is also the *market supply curve*. Analogous to the market demand curve, the market supply curve relates the price of the wildlife good to the quantity supplied to the market by the producers.

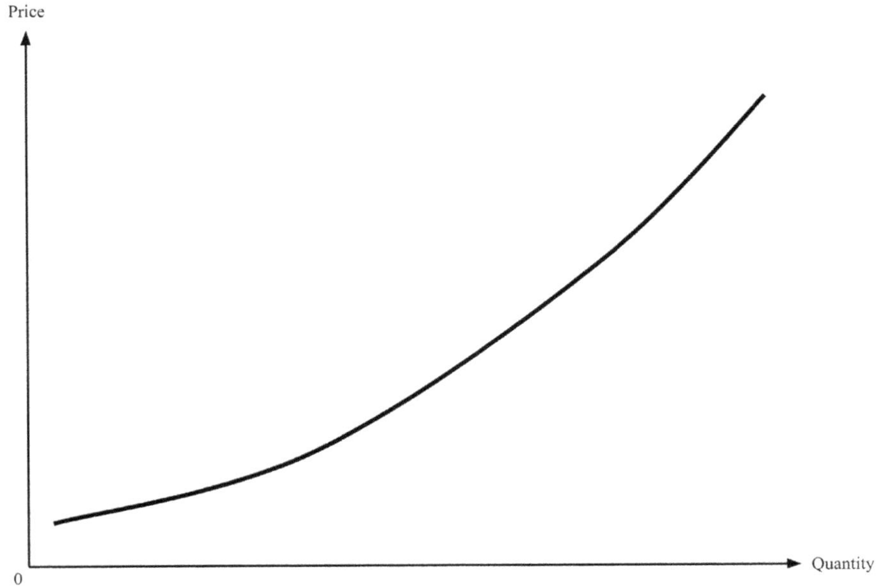

Figure 4.2 A hypothetical marginal cost curve of the suppliers that shows the relationship between the total amount of the wildlife good supplied and the cost of supplying the next unit. If there are no intermediaries in the market so that the producers of the wildlife good sell directly to the consumers, then the marginal cost curve is also the market supply curve if the market is perfectly competitive.

If we are interested in knowing how much the market supply of the wildlife good will be affected by a change in its market price—assuming nothing else changes—we would need to look at the relevant *movement along the market supply curve*. If, however, a factor other than the price of the wildlife good changes, then the impact of this on the supply of the good in the market would be represented by some sort of *shift of the market supply curve*. In this way, the effect on market supply of any changes in any of the determinants of supply—just like for the market demand—can be represented graphically in our economic framework.

Market Equilibrium without Intermediaries

Assuming that there are no intermediaries in the market so that the initial suppliers can sell directly to the consumers, the market would be in *equilibrium*—i.e., in a state of balance where no participants have an incentive to change what they are doing—when the price of the wildlife good is such that the market demand for the good at that price is equal to the amount of good supplied by the market at that price. What that equilibrium price is depends on a host of factors, including how competitive the supply sector is, i.e., what the market structure of the supply side is.[4] We will illustrate this by considering two contrasting scenarios: one in which the supply sector is perfectly competitive[5] and one in which it is monopolized by a single producer.

Equilibrium with Competitive Suppliers

As discussed earlier, when the market is perfectly competitive, the marginal cost curve that shows how the cost of producing the next unit of wildlife good changes with respect to the amount that has already been produced is also the market supply curve. The *competitive equilibrium* of the market can thus be shown graphically (see Figure 4.3) as the point where the market demand curve and the market supply curve intersect each other.

Equilibrium with a Monopoly Supplier

If the supply sector is monopolized by a single producer, then we can no longer use the supply-and-demand model to find the market equilibrium since a monopoly has no supply curve. For this case, the monopoly would need to compare its *marginal revenue*—how much revenue every additional

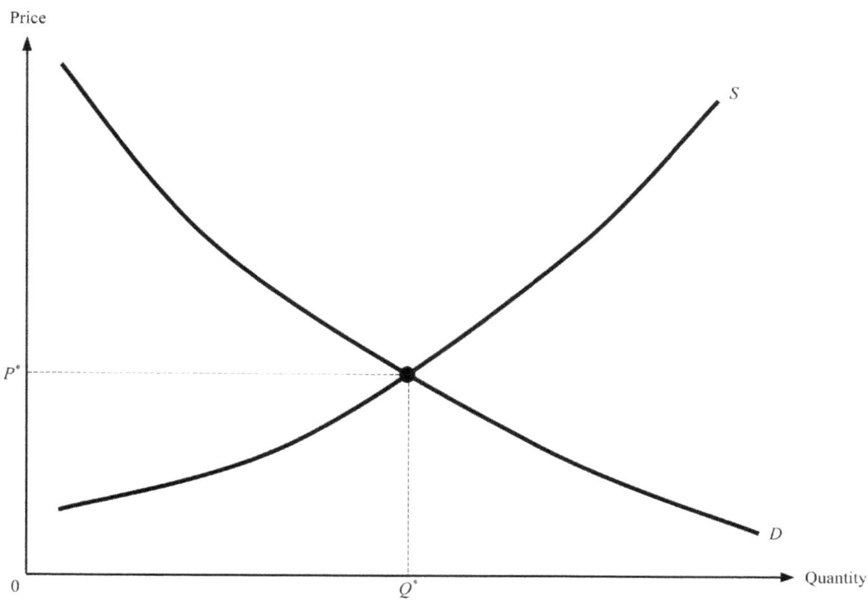

Figure 4.3 When the market is perfectly competitive, the competitive equilibrium of the market is the intersection of the market demand curve (*D*) and the market supply curve (*S*). The equilibrium price is P^*, and the equilibrium quantity is Q^*.

unit of production or supply generates—to its marginal cost of production. This is because, to maximize their profit, the monopolist will keep producing up to the point at which producing one more unit would no longer be profitable. The profit-maximizing quantity for the monopoly—the monopoly quantity—is the one where the monopoly's marginal revenue equals its marginal cost, and the monopoly price is the price at which market demand equals the monopoly output. Figure 4.4 illustrates how the market equilibrium is determined when there is a monopoly supplier of the wildlife good. Note that the monopoly's marginal revenue curve is derived from the market demand curve, and the marginal revenue curve lies below the demand curve. As can be seen from Figures 4.3 and 4.4, if we compare two markets that have the same market demand curve and the same marginal cost curve, the one that is monopolized would have a lower quantity of goods and a higher price compared to the one with perfect competition.

Market Equilibrium with Intermediaries

In practice, markets for wildlife goods only sometimes involve direct transactions between the initial suppliers and the final consumers—as

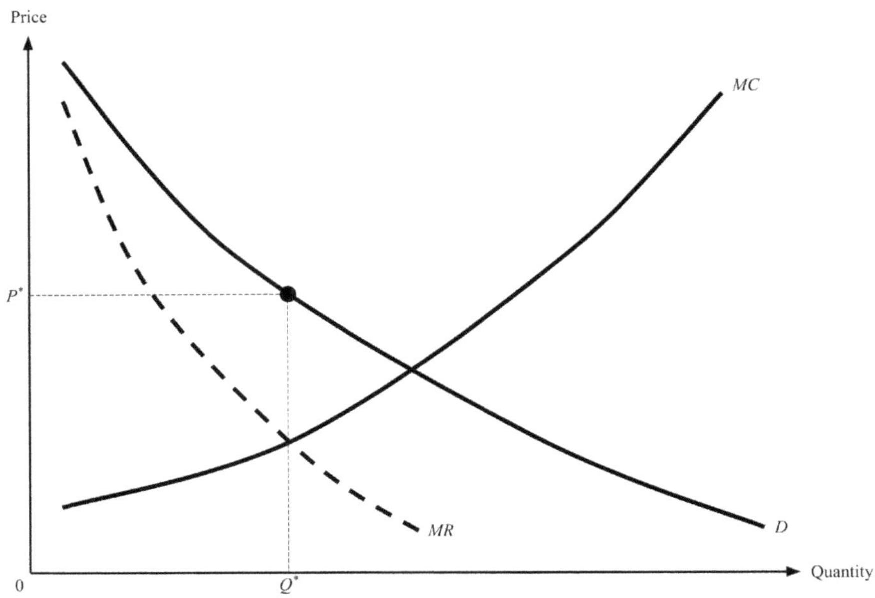

Figure 4.4 When supply is monopolized by a single producer, the monopoly quantity Q^* is the quantity at which marginal revenue (*MR*) equals marginal cost (*MC*). The monopoly price P^* is the price at which market demand equals the monopoly quantity.

discussed in Chapter 3, there are typically intermediaries involved who seek to earn profits through trading and who specialize in the logistics of accumulating, transporting, and distributing products (as well as often processing them along the way). Most wildlife goods markets involve trade chains with multiple intermediaries. In illegal markets, poachers may sell their goods to a local dealer, who accumulates them from several sources before smuggling and/or selling them onward, eventually reaching a wholesale operation, which then sells to several dealers who retail the products to the final consumers. In legal markets the initial suppliers might sell their products to a local wholesaler or export trading operation before the products eventually reach retailers and then consumers. In both instances, intermediaries may be involved in processing or modifying the products to facilitate transport or improve appeal to the final consumer.

For illustrative purposes, let us (for simplicity) assume that there is only one layer of intermediaries between the initial producers and the final consumers of a wildlife good. To highlight the importance of market structure for the operation of the wildlife trade, we will consider two diametrically opposed scenarios: one in which the intermediary sector is perfectly competitive and another one in which it is monopolized by a single entity. Henceforth, we will also assume that the initial producers have little market power and they behave as price-takers.

Competitive Intermediaries

If there is free entry and exit of intermediaries, which are all price-takers and small relative to the whole sector, then their profit—the difference between how much revenue they earn from selling the wildlife good to the final consumers and how much it costs them to procure the good from the producers—would be driven to zero. This is because when the intermediaries are earning a positive profit, the existence of this profit opportunity would entice other intermediaries to enter the market. The increased competition among the intermediaries would then drive down the profit margin. On the other hand, if the profit of the intermediaries is negative, then they would have an incentive to leave the market. These exits reduce competition among the remaining intermediaries, which would, in turn, reduce their losses. Hence, adjustments in the market in response to profit conditions for the intermediaries tend to drive their profit level to zero. Consequently, the price at which the final consumers pay for the good equals the price that the intermediaries pay to the producers to acquire the good.

In addition, for the wildlife market to clear, this common price must be such that the demand from the final consumers at this price equals the amount that the producers are willing to supply to the intermediaries at this price. In this case, the equilibrium with competitive intermediaries would be exactly the same as the market equilibrium without intermediaries in which the sellers are perfectly competitive. Hence, the equilibrium outcome in terms of market price and quantity of wildlife good supplied and demanded would be as illustrated in Figure 4.3.

Monopolistic Intermediary

The situation is significantly different, however, with a monopolistic intermediary that is (1) the only entity that purchases the wildlife good from the initial producers and (2) the only seller of the wildlife good to the final consumers of the product.

In its dealings with the initial producers of the wildlife good, the intermediary would be able to exert a tremendous amount of market power—it has the ability to dictate the price paid to the initial producers—due to its position as the only buyer. It, in essence, would act as a monopsony in purchasing the wildlife goods that are to be sold later to consumers. Its relevant cost curve—i.e., the cost of acquiring the wildlife goods from the initial producers—would be its *marginal expenditure curve* that is derived from the initial producers' marginal cost curve (which is also the supply curve of the wildlife good given our assumption that the initial producers have little market power and they act as price-takers). The marginal expenditure curve has the same price intercept as the supply curve of the wildlife good but is steeper, i.e., the marginal expenditure for the monopoly intermediary rises more quickly with quantity than the marginal cost of supplying the wildlife good for the initial producers (see Figure 4.5).

Being the only seller of the wildlife product to the consumers, the monopoly intermediary can also set the price it sells to them. In determining the profit-maximizing quantity of wildlife good to obtain and then sell, a monopolistic intermediary would thus need to compare the marginal revenue gained from selling an additional unit of the wildlife good to the marginal expenditure incurred from obtaining an additional unit of the wildlife good. The quantity that maximizes the intermediary's profit is the one where its marginal revenue is equal to its marginal expenditure. The market equilibrium with a monopolistic intermediary is illustrated in Figure 4.5. By comparing Figures 4.3, 4.4, and 4.5, we can clearly see that of all the market structures we have considered—a market with no intermediaries and perfect competition; a market with no intermediaries and a monopolistic seller; a market with competitive intermediaries; and a market with a monopolistic intermediary—the market with a monopolistic intermediary has the lowest equilibrium quantity and the highest equilibrium price.

Policy Implications and Applications

We will now apply the economic framework presented in this chapter to consider the effects of certain conservation policies on the equilibrium quantity and price of wildlife goods traded.

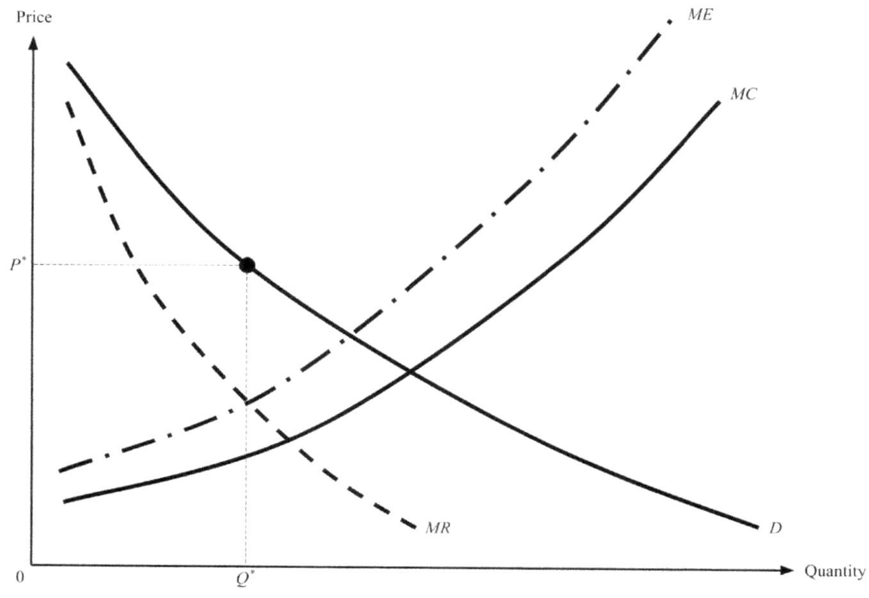

Figure 4.5 When there is a monopolistic intermediary, the equilibrium quantity Q^* is the quantity at which the intermediary's marginal revenue (*MR*) equals its marginal expenditure (*ME*). The equilibrium price P^* is the price at which the market demand equals the equilibrium quantity.

Henceforth, for ease of exposition, we mostly restrict our attention to the situation in which there are no intermediaries in the market. As we saw in the last section, there is no difference—in terms of the equilibrium price and quantity—between a competitive market with no intermediaries and a market with competitive intermediaries. Therefore, we lose nothing in our analysis by concentrating on the market with no intermediaries when we assume that the sellers of the wildlife good—whether they are the initial producers who sell directly to consumers, or the intermediaries who bridge the initial producers and the consumers—are operating under perfect competition.[6] In the final subsection, we briefly consider the policy implications of creating a monopolistic intermediary through regulation, as occurred with two CITES-approved one-off sales of elephant ivory.

Demand-Side Policies

Some conservation policies aimed at curtailing the trade of certain wildlife goods seek to directly reduce consumer demand by lowering the appeal of these products. This is done in various ways, ranging from broad communications to the public to specifically targeting certain consumers; such measures are typically (but not always) also preceded or supported by laws restricting the purchase, consumption, or even ownership of particular wildlife specimens. The measures may include providing consumers with more information about the conservation status of the wildlife species involved; enlisting celebrities to participate in conservation campaigns; highlighting risks associated with consuming those products (e.g., zoonotic disease); and generally attempting to shift social norms relating to the products concerned, i.e., make consumption less socially acceptable.

Analytically, demand reduction policies can be modeled as shifting the market demand curve for wildlife goods downward, indicating that the consumers have a lower willingness to pay (WTP) for the good at every quantity. Policies that are highly effective at changing consumer demand would result in a greater shift of the demand curve, while less efficacious interventions would yield little or no change in the demand curve. Assuming that the market for a wildlife product is perfectly competitive, Figure 4.6 shows that an effective demand reduction policy would result in a lower price for the wildlife good and a lower quantity of it being sold in equilibrium.

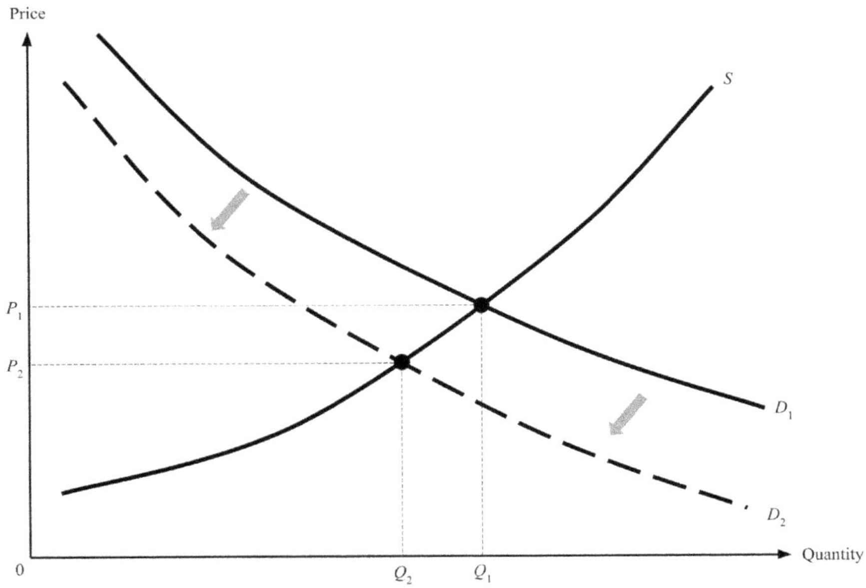

Figure 4.6 A demand reduction policy that shifts the demand curve from D_1 to D_2 reduces the equilibrium price from P_1 to P_2 and the equilibrium quantity from Q_1 to Q_2 in a competitive market.

Supply-Side Policies

To discourage and reduce the amount of trade in wildlife goods, policies could target the supply side of the market as well by changing the producers' or suppliers' incentives to provide and sell the products. Examples of such supply-side policies include trade bans (i.e., making the sales of these goods illegal), any measures to strengthen the enforcement of anti-poaching or anti-trafficking laws, and increasing penalties for breaking anti-poaching or anti-trafficking laws. Another less direct policy approach is to provide potential poachers with more appealing alternative livelihood opportunities.

While the details of implementation, costs, and efficacy of various supply-side conservation policies may differ, they all fundamentally have the effect of increasing the sellers' cost of supplying wildlife goods. In terms of our economic model, these supply-side policies shift the producers' marginal cost curve upward. If the market for wildlife products is perfectly competitive, then—as shown in Figure 4.7—these policies would bring about a higher equilibrium price and a lower equilibrium quantity of the product.

We note that economic policies to enhance the employment or income opportunities of people living near wildlife so that they have less incentive

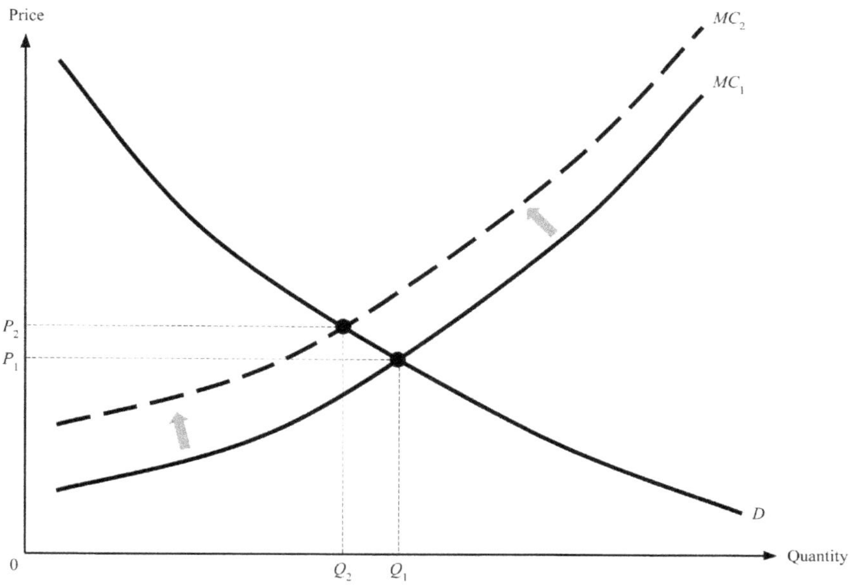

Figure 4.7 A supply-side policy that shifts the producers' marginal cost curve from MC_1 to MC_2 decreases the equilibrium quantity (from Q_1 to Q_2) and increases the equilibrium price (from P_1 to P_2). Note that in a competitive market, the producers' marginal cost curve is the market supply curve.

to turn to poaching to make money have the same effect of shifting the supply curve of wildlife good upward. This is because increasing poachers' (or potential poachers') outside option value—i.e., the attractiveness of income-generating activities outside of poaching—essentially raises their cost of supplying wildlife goods. With better alternatives to poaching, turning to poaching to supply wildlife goods means giving up more in terms of income from these outside options, i.e., the *opportunity cost* of poaching rises. Therefore, the market price of wildlife goods must be higher to induce these people to supply the goods when their outside options are better, and policies that improve their alternatives to poaching result in a lower equilibrium supply of wildlife goods.[7]

Changing the Seizure Rate of Illegal Wildlife Goods

While the analysis given above of the effects of supply-side policies is, as we have seen, fairly straightforward using our economic framework, we should note that the implications derived may be counter-intuitive to some of our readers. Take, for instance, a policy change that increases the seizure rate of illegal wildlife goods, where we define the seizure rate to be the probability that a unit of illegal good offered for sale will be seized. One might conjecture that, since every unit of wildlife good that is produced generates less revenue with the policy change—because that unit becomes more likely to be seized—poachers or suppliers of wildlife goods would want to *increase* the amount of goods they procure or supply in order to make up for the reduced revenue yielded per unit. The implication of this line of thinking is thus that implementing a policy that increases the rate of seizure of illegal wildlife goods could perversely *raise* the amount of wildlife goods that are bought and sold in the market.

This conclusion, however, does not follow from the economic framework presented here. To see this, first, note that increasing the seizure rate of wildlife goods in our economic model is equivalent to increasing suppliers' marginal cost of production or supply. Let us consider why this is so.

A profit-maximizing seller, when calculating whether to supply a unit of good or not—or if they have already decided to supply at least one unit, whether to supply one more unit or not—will compare the revenue expected to be generated from supplying that unit to the cost of supplying that unit—the latter of which is by definition the seller's marginal cost of supplying that unit. If we denote by s the probability with which a unit of wildlife good supplied by the seller will be seized by authorities, then the revenue that the seller can expect from supplying that unit is $(1-s)P$, where P is the market price of the product. If $(1-s)P > MC$, where MC denotes the marginal cost of the unit, then it would be profitable for the seller to supply the unit; otherwise, there is no net gain to the seller from supplying the unit.

Therefore, the seller's supply decision involves comparing a unit's expected revenue, $(1-s)P$, with its marginal cost, MC. Crucially, this comparison of expected revenue and marginal cost is *completely equivalent* to the comparison between the product price P and $\frac{MC}{1-s}$, which is the unit's marginal cost scaled by the probability that it will not be seized.

The above observation tells us that, when there is a non-zero probability that every unit of wildlife good will be seized by authorities, the relevant marginal cost curve we need to look at—which gives us the market supply curve of wildlife goods—is not the actual marginal cost curve of the suppliers; rather, it is the curve we obtain after multiplying the actual marginal costs by the factor $\frac{1}{1-s}$. This scaling factor allows us to account for the seizure rate of the goods.

Notice that as the seizure rate increases—i.e., as s rises—the scaled marginal costs go up. Therefore, increasing the seizure rate in our economic model is represented by a shift up (or to the left) of the market supply curve of wildlife goods. As we have seen earlier, such a shift has the effect of *reducing* the equilibrium quantity of wildlife goods supplied and increasing the market price.

The Snob Effect

One of the forces that contribute to and exacerbate conservation issues is the fact that many consumers place a higher value—have a greater WTP—for wildlife products that are rare or uncommon (Gault et al., 2008; Angulo et al., 2009; Johnson et al., 2010; Lyons & Natusch, 2013). This is what economists refer to as the *snob effect* (Leibenstein, 1950). In our economic framework, this means that the market demand curve for a wildlife good is higher the less of it is available on the market.

One major policy implication of the snob effect in the context of wildlife trade is that its presence could significantly reduce the effectiveness of supply-side conservation policies in decreasing the amount of trade. To see this, consider a competitive market for some wildlife product and a supply-side policy that shifts the market supply curve inward. Absent the snob effect, the market demand curve remains as it is after the implementation of the policy, and the market equilibrium changes, as shown in Figure 4.7.

With the snob effect, however, the initial reduction in trade wrought by the policy would cause an increase in consumer demand, which results in an upward shift of the market demand curve. As shown in Figure 4.8, this change in consumer valuation would thus partly counteract the initial effect of the supply-side policy. While the snob effect cannot fully or more than offset the supply-side policy in terms of quantity of wildlife good traded (Chen, 2016), its impact—which affects the effectiveness of the supply-side policy—could be large depending on how strong the effect is.

Cost of Policy Implementation and Changes

As the discussions and analyses above show, various demand- and supply-side policies could be effective in curbing the trade in wildlife goods and reducing

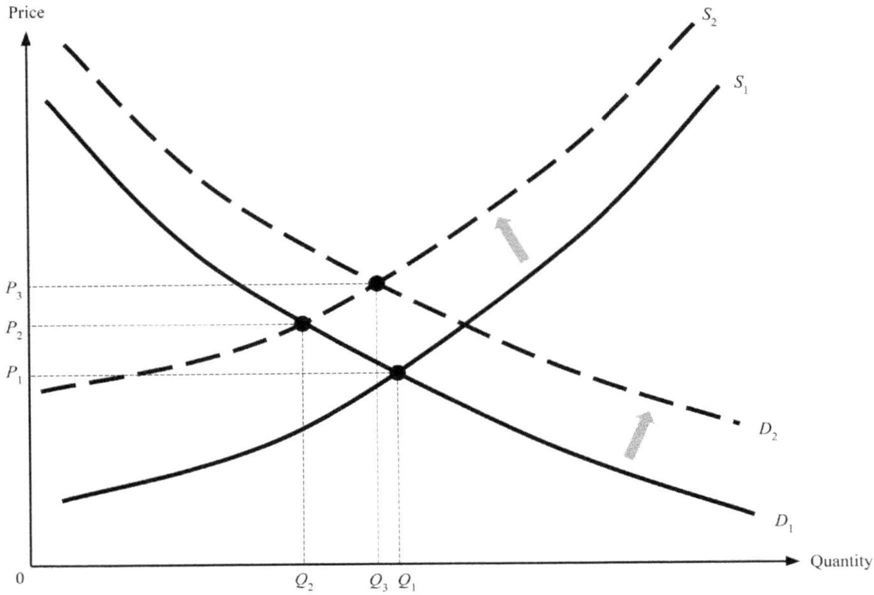

Figure 4.8 In a competitive market, a supply-side policy that shifts the supply curve from S_1 to S_2 increases the market price from P_1 to P_2 and decreases the market quantity from Q_1 to Q_2 in the absence of the snob effect. With the snob effect, however, the initial decrease in trade brought about by the supply-side policy shifts the demand curve from D_1 to D_2. Taking into account the snob effect, the policy therefore increases price from P_1 to P_3 and decreases quantity from Q_1 to Q_3.

their equilibrium supply in the market. We emphasize, however, that different types of demand-side policies could vary greatly in their cost of implementation and in their efficacy in discouraging the trade in wildlife goods. The same point applies to supply-side policies. Note also that some conservation policies—e.g., a fine for poachers who are caught and penalized for their actions—could even yield law enforcement agencies or the governments implementing them a revenue stream. Hence, a "holistic" evaluation of a conservation policy or measure needs to consider all the benefits that could be expected to result from the policy, as well as all the costs that could be incurred from its implementation. An in-depth examination of this issue is beyond the scope of this book since our central focus here is to explicate how various factors or policies can affect the equilibrium of the market for wildlife goods, rather than a cost–benefit analysis of various conservation instruments or policies.

Legal versus Illegal Wildlife Good

We stated at the beginning of this chapter that the exposition in this chapter assumes that we are looking at a single wildlife good or product and that the entire market is either legal or illegal. Now, it is possible that a wildlife good comes in different versions or varieties; on top of that, legal status may vary—for example: trade in rhino horns, and wild lion skulls may be legal if they are part of a certified safari hunting trophy, but not otherwise; from 2017 export trade of skeletons from captive-bred lions in South Africa was legal (subject to a quota), but trade in wild-harvested bones was not[8]; similarly, commercial trade in elephant ivory is illegal between all countries and also banned domestically in many countries, but is still legal within several others. In such cases, the different versions of the wildlife good are substitutes for one another, and a proper analysis of the trade in them must take into account the supply of and demand for all the versions simultaneously. In other words, for these instances, we cannot look at the trade in a single variety or version of the good in isolation from the trades in other versions. To see how the economic model works for these cases—which are considerably more complicated than the case analyzed in this chapter—the next two chapters, Chapters 5 and 6, consider how markets for a wildlife good work when the product comes in two versions: a wild-sourced variety, and a substitute variety (which may come from farmed animals or which may be a human-made synthetic substitute).

Before moving on to Chapter 5, however, it is important and instructive to consider briefly—when only a single version of a wildlife good is available—how the analysis of the market for the good is affected by the legal status of the trade in the good. Let us suppose for the sake of argument that trade of the wildlife good in question is initially illegal, and it becomes legalized subsequently. Figure 4.9 shows the supply and demand curves in this market for the two cases.

When trade becomes legal, the cost of supplying an additional unit of the wildlife good for sellers decreases as it becomes easier in some sense to supply the good. Hence, the marginal cost of production or supply decreases, which is indicated by the downward shift of the supply curve. At the same time, on the buyers' side, the stigma of purchasing the good diminishes once trade becomes legalized, and it becomes easier for the buyers to make purchases once transactions no longer need to be covert. These effects are reflected graphically by an upward shift of the market demand curve.

As Figure 4.9 shows, the shifts of the supply and demand curves unambiguously increase the equilibrium amount of wildlife good traded. However, the impact of trade legalization on the equilibrium price is ambiguous. While the downward shift of the supply curve tends to decrease the equilibrium price, the upward movement of the demand curve puts upward pressure on the price. Therefore, what ultimately will happen to the price in equilibrium with legal trade depends on whether the impact on the supply curve is larger or smaller than the impact on the demand curve.

Monopolistic Intermediary

As a final example of a policy-relevant issue, we consider a real-life case study of a monopolistic intermediary and its implications. In 1989, the Parties to CITES

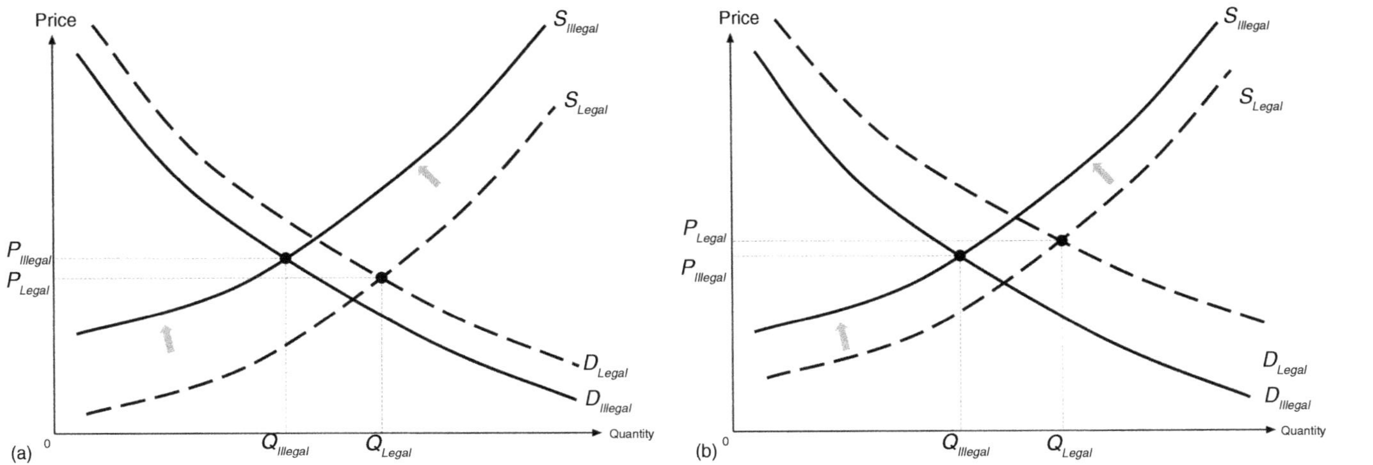

Figure 4.9 Legalizing trade in the wildlife good increases the equilibrium quantity of the good but has an ambiguous effect on the equilibrium price. (a) The effect on the supply curve is relatively stronger than the effect on the demand curve, and the equilibrium price decreases. (b) The effect on the supply curve is relatively weaker than the effect on the demand curve, and the equilibrium price increases.

agreed on an effective international ban on elephant ivory trade. However, several southern African range states (with abundant elephant populations) objected to this and started to lobby for the resumption of a controlled legal trade between certain countries. As a result, in 1997, the CITES Parties agreed to allow a tightly controlled "one-off" government-to-government sale between three range states and one consumer country, Japan. By doing this, they gave Japan effective monopoly buying power for the sale, which took place in 1999. A second CITES-approved one-off sale took place in 2008, between four African range states and two consumer countries (Japan and China). Despite the potential for competition between the two buying countries, they colluded to again act as a monopoly buyer (i.e., a monopsony). In China's case, the buying entity then retained most of the stockpile, only releasing small amounts of ivory onto the market afterward, thus effectively behaving like a monopolistic intermediary as described earlier in this chapter.

Why did CITES enable this monopolistic intermediary market structure, and what was the outcome? The CITES Parties are greatly concerned about controlling the market and preventing the introduction of illegally obtained (i.e., poached) ivory from entering legal market channels—so-called "laundering" (as discussed in Chapters 1 and 3)—and believe that restricting market participation can help to prevent this. The outcome in terms of pricing was much as predicted by our basic model, with the monopolistic intermediaries achieving a lower buying price than the known illegal market value; similarly, the small amounts of stock then released to the market were at significantly high prices, such that the intermediary took a large profit.

Is this one-off sale policy a good or bad thing for conservation? There are two issues to consider here. The first is that the initial suppliers (the elephant range states) received less money for their ivory than they would have with true competition among the buyers. Because at least some of this income is used to finance conservation, having it reduced in this way is likely not a good conservation outcome. The second issue to consider is the monopoly power that the buying entity possesses when it resells some ivory from its stock to the consumers. The high price that the buying entity can command in the consumer market could induce entry of competing illegal suppliers into the market, which would be a detriment from a conservation standpoint. A full analysis of this issue requires an examination of how a legal market for a wildlife good is affected by a parallel illegal market. In the following two chapters, we expand on our basic economic model and consider how a wildlife goods market operates when multiple versions of a good are available. These more general frameworks allow us to consider, among other applications, how trade in wildlife goods is impacted when a legal market and an illegal market exist simultaneously.

Lessons for Policy and Practice

Illustrated by several examples, we have demonstrated how our basic economic model provides an understanding of market mechanisms and how they might influence the quantities of wildlife goods being traded, with consequent conservation implications (i.e., higher trade volumes imply potentially higher harvesting rates). By drawing and examining the supply and demand curves for any transaction in a wildlife product trade chain, we can analyze how changes in prices would affect quantities traded and how shifts in the supply and demand curves might affect both prices and quantities. In Chapter 8, we adopt a somewhat different but complementary approach when we examine the incentives of specific actors in the trade chain as another way to consider the likely effects of various policies and practices. In the meantime, however, we build upon our basic supply and demand model to consider the effects of substitute products in the following two chapters.

Notes

1 Note that some of these products have an added element of human craftwork, which enhances their final economic value. However, for the purposes of our simple model it may be easier to think about a generic unmodified wildlife good.

2 It is often assumed that simple "stricter" measures—such as increasing the penalty for poaching—alone will raise the expected cost of poaching, but this is not necessarily true if the probability of incurring such a penalty is varied as well. For example, in some cases raising penalties can lead to a reduced probability of incurring them because enforcers may be more careful to inflict harsher penalties. Hence, we refer to "more effective" measures, being those that do in fact raise the expected cost of poaching; one such measure is to increase the probability of early interception of poachers (before they succeed in extracting a wild specimen).

3 The concepts of market power and perfect competition are explained further in the subsequent endnotes.

4 By "market structure" we mean the number of buyers and sellers and the extent to which one or more of them may have some sort of advantage that allows them to access and control a disproportionate part of that market, such that they could achieve lower (for buyers) or higher (for sellers) prices than the average. This ability to exert such influence over prices is typically referred to as "market power," and its ultimate expression is a monopoly (when there is only one seller) or monopsony (when there is only one buyer), a situation that enables complete control over price-setting.

5 For our purposes here, we consider a market to be perfectly competitive when no buyers or sellers can individually exert any control over the market, i.e., they have no market power. In this setting, all buyers and sellers are price-takers—because they have no ability to affect market price on their own, they take the price as given when deciding whether or how much to buy (in the case of buyers) or sell (in the case of sellers).

6 While the assumption that sellers of wildlife goods are price-takers may be overly restrictive, we note that the analysis of wildlife trade with imperfect competition, with a few exceptions later on in the book, lies outside the scope of this book. This is because analysis of imperfect competition requires more than the basic supply-and-demand model of economics, and this book presupposes that the readers do not have knowledge of economics beyond the supply-and-demand framework.

7 Note that we are assuming here that poaching is economically motivated. Some empirical research has shown that illegal hunting can also be driven by other (e.g., cultural, recreational) motivations (Knapp, Peace, & Bechtel, 2017) and that in such cases increasing income can have perverse effects such as enabling the purchase of better hunting equipment or providing more leisure time in which to hunt (Damania et al., 2005).

8 This situation changed after the quota was suspended due to protests—this was an interesting, complicated, and controversial case study, as discussed by Williams and 't Sas-Rolfes (2019).

References

Angulo, E., Deves, A. L., Saint Jalmes, M., & Courchamp, F. (2009). Fatal attraction: Rare species in the spotlight. *Proceedings of the Royal Society B: Biological Sciences, 276*(1660), 1331–1337. https://doi.org/10.1098/rspb.2008.1475

Chen, F. (2016). Poachers and snobs: Demand for rarity and the effects of antipoaching policies. *Conservation Letters, 9*(1), 65–69. https://doi.org/10.1111/conl.12181

Damania, R., Milner-Gulland, E. J., & Crookes, D. J. (2005). A bioeconomic analysis of bushmeat hunting. *Proceedings: Biological Sciences, 272*(1560), 259–266.

Gault, A., Meinard, Y., & Courchamp, F. (2008). Consumers' taste for rarity drives sturgeons to extinction. *Conservation Letters, 1*(5), 199–207. https://doi.org/10.1111/j.1755-263X.2008.00038.x

Johnson, P. J., Kansky, R., Loveridge, A. J., & Macdonald, D. W. (2010). Size, rarity and charisma: valuing African wildlife trophies. *PLoS One, 5*(9), e12866.

Knapp, E. J., Peace, N., & Bechtel, L. (2017). Poachers and poverty: Assessing objective and subjective measures of poverty among illegal hunters outside Ruaha National Park, Tanzania. *Conservation and Society, 15*(1), 24–32. https://doi.org/10.1371/journal.pone.0012866

Leibenstein, H. (1950). Bandwagon, snob, and Veblen effects in the theory of consumers' demand. *The Quarterly Journal of Economics, 64*(2), 183–207. https://doi.org/10.2307/1882692

Lyons, J. A., & Natusch, D. J. (2013). Effects of consumer preferences for rarity on the harvest of wild populations within a species. *Ecological Economics, 93*, 278–283. https://doi.org/10.1016/j.ecolecon.2013.06.004

Williams, V. L., & 't Sas-Rolfes, M. J. (2019). Born captive: A survey of the lion breeding, keeping and hunting industries in South Africa. *PLoS One, 14*(5), e0217409. https://doi.org/10.1371/journal.pone.0217409

5 The Economic Model with Substitute Goods and Complete Information

Introduction and Overview

For certain wildlife goods, suitable alternatives or substitutes may be or potentially be available for the consumers to purchase. Take, for example, the trade in tiger bones, or more specifically the trade in bones sourced from wild tigers. Bones obtained from farmed tigers can act as a substitute for wild tiger bones. Indeed, for many wildlife goods, farmed substitutes may readily be available. For other wildlife goods—rhino horns, for instance—the possibility of engineering and manufacturing high-fidelity synthetic substitutes that are biochemically identical to the real, natural goods has been discussed in the news as a potential solution to the problem of wildlife poaching.

In this chapter, we examine the potential impacts of introducing a substitute good into a market for a wildlife good. A critical consideration when analyzing the effects of the availability of a substitute good on the supply or consumption of a wild good is the extent to which consumers can distinguish between the two types of goods. This matters for the following reasons. First, in many cases, consumers consider farmed or synthetic goods to be imperfect substitutes for wild goods—they value substitute products lower than their wild counterparts.[1] Second, if consumers have the ability—or are provided with the necessary information—to differentiate between substitute goods and wild goods, then there would actually be two separate markets—each with their own price—for the two types of goods. In addition, when wild goods are generally preferred, we would expect the price for the substitute good to be lower than that for wild good to reflect the consumers' differing valuations for the two products.

In contrast, if consumers are unable to distinguish between a wild good and a substitute good, then only one market would effectively exist—where both types of goods are lumped together—with only one price. This is so because, without the ability to differentiate between the different types of goods, the price that consumers pay cannot be made dependent on whether or not a good is obtained from the wild.

For this chapter, we will focus on the case in which there is complete information in the market so that consumers can perfectly differentiate between

DOI: 10.4324/9781003286318-5

the wild product and the substitute product (more precisely, consumers can perfectly differentiate between the two types of goods without incurring any cost). This complete information may come about for different reasons. For instance, the two types of goods—despite their similarities—may have enough differences for consumers to tell them apart. The case of wild salmon and farmed salmon fits this scenario since the two types of salmon are fairly easy to distinguish visually.

Another reason for why there could be complete information in the market—when the substitute good is virtually indistinguishable from the wild good—is the existence of a credible certification mechanism that is designed to inform the consumers. For example, if trade in the wild good is illegal, but trade in the substitute good is legal, a trustworthy certification system that labels only the substitute good as such allows consumers to differentiate between the two types of goods. Such a certification system may be voluntary or may be further supported by regulation, through a permit system, the existence of which can also provide a way to signal information. There are many instances in which governments only allow domestic wildlife trade to take place after a relevant authority has issued a permit. Permits also provide the foundation for the CITES system of international trade regulation.

In this chapter, we take as our starting point in the analysis that there is complete information in the market without addressing how or why this complete information came about. This means that if the complete information is the result of having a credible certification system in place, we do not consider the specifics of the certification system or how its credibility is maintained; nor do we consider the incentives of the sellers to obtain or not obtain such certification. We leave these issues—including the issue of sellers of illegal wild goods having an incentive to use a legal trade in a substitute good to launder their products—to the following chapter to examine. Note that all derivations and proofs of the results presented in this chapter are given in the technical appendix (available at: https://www.frederickchen.org/publications/the-economics-of-wildlife-trade-technical-appendices).

The Modeling Framework

Demand

Given two alternatives—a preferred wild good and a substitute product—to choose from, a consumer's type or characteristics can be captured in our model by two numbers: $V_w \geq 0$, which denotes a consumer's valuation of or willingness to pay for the wild good, and $V_s \geq 0$, which denotes a consumer's valuation of the substitute good. As discussed previously, it should be the case that $V_w \geq V_s$.[2] For illustrative purposes, a consumer's type can thus be represented simply as a point in the shaded region in Figure 5.1, where the *x*-coordinate gives the consumer's V_w, and the *y*-coordinate is the consumer's V_s. Since consumers may differ from one another in terms of how much they are willing to pay for the two types of goods, one consumer's "location" in this (V_w, V_s)-space may be distinct from another's location. Therefore, we would expect, in general, for the consumers to be scattered throughout the shaded region shown in Figure 5.1.

Since buyers can distinguish between the wild good and the substitute good by assumption in this chapter, there are two prices we need to consider: P_w, the price of the wild good; and P_s, the price of the substitute good. For simplicity, we will assume that each buyer can buy only one unit of good.[3] Given prices P_w and P_s, we can derive the demand for the wild good (D_w) and the substitute good (D_s) as follows. The net benefit (or *consumer surplus*, in economics parlance) from buying the wild good is $V_w - P_w$, and the consumer surplus from buying the substitute good is $V_s - P_s$. It is optimal for a buyer of type (V_w, V_s) to choose the wild good if the consumer surplus

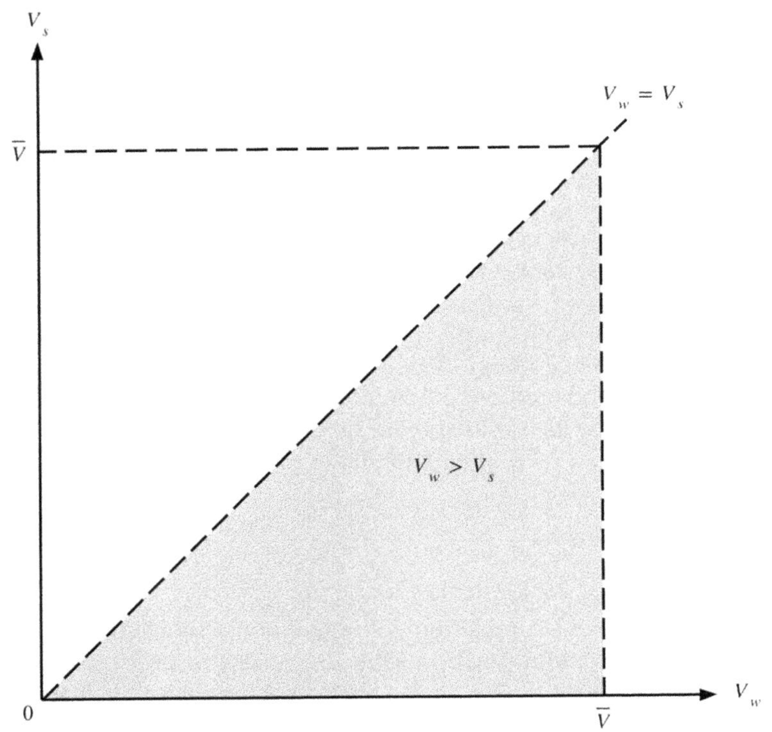

Figure 5.1 A consumer's type can be represented as a point in the (V_w, V_s)-plane, where V_w and V_s, respectively, are the consumer's willingness to pay for the wild good and the substitute good. Since farmed or synthetic goods are imperfect substitutes for the wild goods, we have $V_w \geq V_s$, and only the shaded region in the plane need to be considered. $\overline{V}$ denotes the highest value that V_w or V_s can take.

from buying the wild good is positive and exceeds the consumer surplus from buying the substitute good, i.e., $V_w - P_w \geq V_s - P_s$ and $V_w \geq P_w$; and the buyer would opt for the substitute good if the consumer surplus from purchasing it is positive and greater than the consumer surplus from buying the wild good, i.e., $V_w - P_w \leq V_s - P_s$ and $V_s \geq P_s$. If $V_w - P_w = V_s - P_s \geq 0$, the buyer would be indifferent between the two goods and would be willing to choose either one. Figure 5.2 shows graphically using the (V_w, V_s)-space the demand for the wild good and the substitute good given prices P_w and P_s.

Some properties of the demand for the wild good and the demand for the substitute good—which are very intuitive—can be seen clearly from Figure 5.2. All else equal, an increase in the price of the substitute good reduces the demand for the substitute good (the area corresponding to D_s shrinks) and raises the demand for the wild good (the area corresponding to D_w expands). On the other hand, an increase in the price of the wild good reduces the demand for the wild good (the area corresponding to D_w shrinks) and increases the demand for the substitute good (the area corresponding to D_s expands).

Supply

Given complete information on the part of buyers such that they can perfectly distinguish between the wild good and the substitute good, and assuming that both the market for the wild good and the market for the substitute good are perfectly competitive, there would be a market supply curve for the wild good and a separate market supply curve for the substitute good. Assuming profit maximizing behavior by all sellers, these market supply curves would simply be the marginal cost curves of the sellers.[4] As noted in Chapter 4, all these marginal cost curves should be upward sloping, i.e., the higher the quantity of production or supply, the higher the marginal cost. Hence, the supply curves for the wild good and the substitute good should slope up—the higher the price of a good, the greater the quantity supplied by sellers of the good.

For the rest of our exposition, we assume that the substitute good is cheaper to supply than the wild good in the following sense: the marginal cost curve of the substitute good lies below the marginal cost curve of the wild good at quantity 0. Intuitively, this means that the cost of supplying the first unit of the substitute good is lower than the cost of supplying the first unit of the wild good. We place no additional restrictions on the marginal cost curves other than that the two curves should be upward sloping throughout.

We note that the assumption that the marginal cost curve for the substitute good starts lower than the marginal cost curve for the wild good is not crucial for the results that follow. If we assume that the marginal cost curve for the substitute good lies higher than the marginal cost curve for the wild good, then two possible cases could arise in equilibrium: a trivial case in which the price of the substitute good is greater than the price of the wild good, and no one supplies or purchases the substitute good; and the non-trivial case in which the price of the substitute good is less than the wild-good price, and

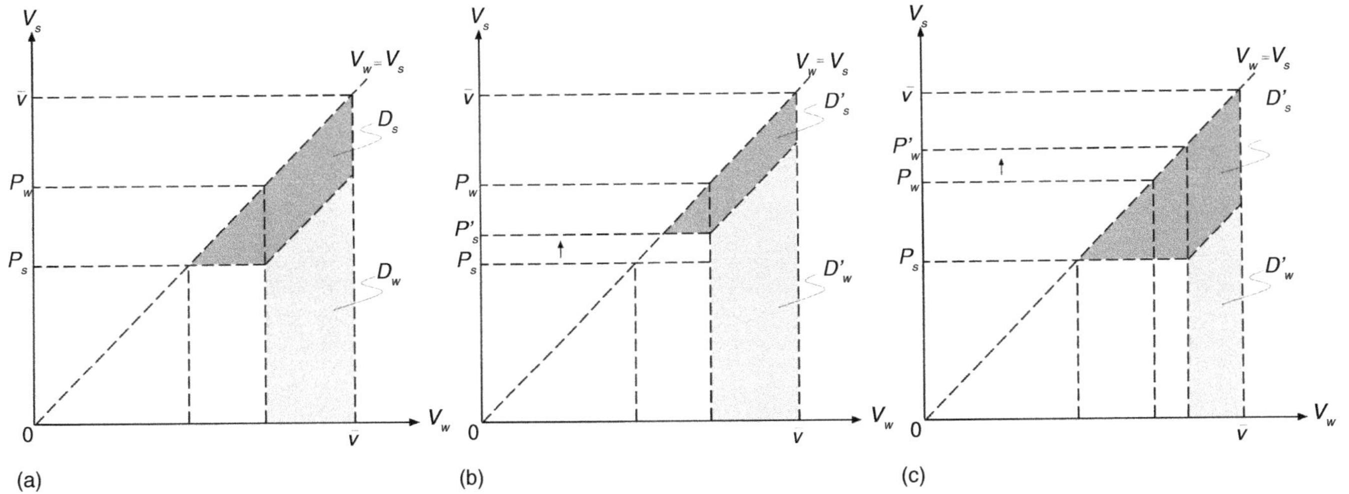

Figure 5.2 (a) Given prices P_w for the wild good and P_s for the substitute good, the demand for the wild good is given by the number of consumers with valuations in the area denoted by D_w, and the demand for the substitute good is given by the number of consumers with valuations located in the area marked D_s. (b) When the price of the substitute good rises from P_s to P_s' (and the price of the wild good stays the same), the demand for the wild good increases ($D_w' > D_w$), while the demand for the substitute good falls ($D_s' < D_s$). (c) When the price of the wild good increases from P_w to P_w' (and the price of the substitute good remains the same), the demand for the wild good decreases ($D_w' < D_w$), while the demand for the substitute good increases ($D_s' > D_s$).

positive quantities of the substitute good are bought and sold. Our results below apply with no modifications in the non-trivial case.

Market Equilibrium

The markets for the wild good and the substitute good are in equilibrium when they both clear, i.e., when the supply of a good equals the demand for it in each market. Since the price of the wild good affects the supply of the good, as well as the demand for both the wild good and the substitute good, and since the price of the substitute good affects its supply, in addition to the demand for both the wild good and the substitute good, the two markets can reach equilibrium only if both prices, P_w and P_s, are "just right."

Properties of the Market Equilibrium

To examine how various conservation policies could affect the supply of or demand for the wild good and the substitute good—and in turn the equilibrium in both markets—let us first consider some key properties of the market equilibrium.

- The market equilibrium must be unique in the sense that there is only one pair of prices, (P_w, P_s), that clear both the market for the wild good and the market for the substitute good. This point is notable because, as we shall see in the next chapter, when consumers are not able to distinguish between the wild good and the substitute good, multiple market equilibria that differ in the amount of goods supplied could arise.
- Given our assumptions that (1) buyers value the wild good more than the substitute good and that (2) the substitute good is cheaper to supply than the wild good (at least for the first unit), the price of the wild good must be higher than the price of the substitute good in equilibrium.[5]
- If we compare the equilibrium price of the wild good when the market for the substitute good does not exist to the equilibrium price of the wild good when there is a market for the substitute good, the price of the wild good when there is no market for the substitute good has to be higher than the price of the wild good when the substitute good is available for sale. In other words, the presence or existence of the substitute-good market drives down the price of the wild good.

Policy Implications

Let us now apply our economic model to consider some policy implications.

Effects of Subsidizing the Substitute-Good Sector

Suppose the sellers or producers of the substitute good are subsidized by entities such as government agencies or NGOs for producing their wares. This has the effect of reducing the marginal cost of producing the substitute good. In a competitive market, this leads to a downward shift of the market supply curve of the substitute good. As a result, the price of the substitute good decreases. This, in turn, takes away some of the demand for the wild good, leading to downward pressure on the price of the wild good as well. Hence, both the price of the wild good and the price of the substitute good are lower in equilibrium when the substitute-good sector is subsidized (compared to the situation in which it is not subsidized). This, in turn, leads to a lower equilibrium quantity for the supply (and consumption) of the wild good.

Effects of Stronger Anti-Poaching Measures

Now, let us consider the impact on the markets for the wild good and the substitute good if more effective anti-poaching or other supply-restrictive measures were implemented. Whatever form these measures or policies take, most of them have the effect of increasing the (marginal) cost of supplying or selling the wild good. In terms of our economic framework or model, this is represented by an upward or inward shift of the wild-good supply curve. This puts upward pressure on the price of the wild good, which subsequently causes the price of the substitute good to go up as well since a higher price for the wild good causes some buyers to shift their demand from the wild good to the substitute good. In the equilibrium with tougher anti-poaching policies, therefore, both the wild good and the substitute good are more expensive to purchase. More importantly from a conservation perspective, the equilibrium quantity of the wild good sold and purchased is lower relative to the situation with less effective or absent anti-poaching policies.

Effects of Demand-Side Promotion of the Substitute Good

Suppose, in order to curb poaching and the demand for the wild good, policies are implemented to promote the purchase of the substitute good.[6] Such policies could be modeled in our framework as causing individual buyers' valuation of the substitute good, V_s, to go up (without affecting the valuation of the wild good, V_w). They would have two opposing effects on the equilibrium price of the wild good. All else fixed, demand-side policies promoting the substitute good reduce the demand for the wild good by causing consumers to shift away from purchasing the wild good toward buying the substitute good. This puts downward pressure on the price of the wild good, and is a direct effect of the demand-side policies.

Not surprisingly, using demand policies to promote the purchase of the substitute good would increase the equilibrium price of the substitute good, causing its market supply to rise. In turn, the rise in the price of the substitute good, all else fixed, would result in an increase in the demand for the wild good. This can be thought of as an indirect effect of the demand-promotion policies, and it tends to increase the equilibrium price of the wild good.

The direct and indirect effects of the demand policies on the price of the wild good act in opposite directions, and ultimately what happens to the equilibrium price of the wild good depends on their relative magnitudes. In particular, it is important to note that it is indeed possible for conservation policies aimed at stimulating demand for the substitute good to perversely *increase the price of the wild good and thereby raise the supply of the wild good in equilibrium*. We demonstrate this possibility in the technical appendix by using a numerical example of our economic model. This example underscores the importance of understanding the details of the market—e.g., the preferences of consumers for the different types of goods and the cost structure of the suppliers—in designing effective conservation policies.

Contested Policies

To conclude this chapter, we consider some recent discussions in the literature and among policymakers about the existing and potential role of trade in substitute products supplied by wildlife farming, manufacturing using synthetic biology, and other natural "look-alike" products, which we call "simulants." Having reviewed this literature, we find some confusion in the discussions, potentially leading to prescriptions that do not align with our preceding analysis. We highlight some of the inconsistencies in this section.

Wildlife Farming

As mentioned in Chapter 2, wildlife farming and associated trade are rapidly expanding to meet the growing consumer demand for wildlife products. This trend elicits concerns both from conventional conservationists, who worry about potential adverse impacts on wild species populations, and from people who care about animal welfare, who worry about harm to individual animals raised in captivity. Since the COVID-19 pandemic, farming of wild animals also increasingly raises concerns about its potential to elevate zoonotic disease spillover risk. These growing concerns have placed certain wildlife farming practices in the spotlight and prompted resistance in countries such as South Africa, in which a policy review in the early 2020s sought to place limitations on the further expansion of the wildlife farming sector, even reversing previous developments—notably those of commercial captive lion breeding and rhino farming.

Several researchers have attempted to develop frameworks for assessing conditions under which wildlife farming can be beneficial for conservation (e.g., Phelps et al., 2014; Tensen, 2016; Meeks et al., 2024). However, such frameworks typically offer assessments based on unweighted (and thus somewhat random) checklists or decision-tree-type approaches that ignore some critical economic insights. Our analysis shows that in cases where farmed and wild products are readily distinguishable by consumers, the sale of cheaper farmed products, even if considered an inferior substitute, may help to depress the prices and quantities of wild products sold. This is because the sale of a competing (even if imperfect) substitute displaces some of the consumer demand for the wild product. Some framework assessments mistakenly imply that farming is inadvisable as a conservation measure if consumers prefer products harvested from the wild (e.g., Tensen, 2016: 288, Table 1).

Assessments of wildlife farming also typically ignore the critical difference between policies that effectively subsidize the farmed product (thereby making it even more price-competitive than the wild product) and those that actively promote the farmed product in order to grow consumer demand (and therefore raise selling prices), which may have a negative spillover effect on the wild product. It is worth noting that allowing legal trade in a farmed product while banning sales of the wild product amounts to an effective form of subsidy for the farmed product and will tend to suppress sales of the wild product as long as the two products can be readily distinguished or kept separate. Following the same logic, subsequently banning sales of the distinguishable farmed product, we might expect two consequences. First, if any demand for the farmed product persists, some (illegal) trade in it will continue, albeit at lower quantities, given that the farmed product has now become relatively less competitive with the wild product. Second, and more consequential for conservation, the increased competitiveness of the wild product will boost its relative sales and thus lead to increased poaching to supply the overall market.

The current framework assessments tend to miss the nuances of marginal effects, as described above, and rather present choices as absolute. As an example, Tensen (2016) proposes one criterion as "no restocking from the wild," implying that any future flows of wild live specimens into the farming breeding stock whatsoever will negate any conservation benefit. This is clearly incorrect, as some harvesting from the wild to boost the quality of the farmed population could have positive effects for both, provided that it takes place at sustainable levels of extraction from the wild. The positive effect for farming could be to increase the genetic robustness of its population and lower production costs (an effective subsidy), and the positive effect for the wild population would result from the process described in the previous paragraph.

These analyses raise another important point that tends to be overlooked in the framework assessments, i.e., that every case should be examined based on current circumstances when considering a new policy recommendation. Introducing a novel farmed product to a situation in which it does not yet exist is not equivalent to intervening in a market in which farmed products are already being sold. Several critiques of existing wildlife farming practices argue from hypothetical perspectives rather than an assessment of current market conditions and how they may be impacted by policy change.

As a final caveat, we reiterate that the analyses in this chapter apply to conditions in which separate supply channels for wild and farmed products are easily maintained, enabling them to be readily identified as different products. Under circumstances where they are not easily separated or identified as different, the conclusions here may be undermined, as discussed in the following chapter. This is due to the potential for "laundering" and successful fraudulent representation, as explained in Chapters 1 and 3. These concerns apply to both farmed products and the other types of potential substitutes that we now turn to discuss.

Synthetics and Distinguishable Simulants

Another category of products of concern to conservation policymakers relates to manufactured or other natural products that resemble the products of threatened species. Examples would include products manufactured using technologies such as synthetic biology (e.g., lab-grown meat), distinct products from other species that are obviously fraudulently marketed (e.g., buffalo horn labeled as rhino horn), and products from extinct species that resemble those of extant species (e.g., mammoth ivory). While largely indistinguishable substitutes raise the concerns highlighted above and discussed in Chapter 6, those that are readily distinguished and marketed separately may provide positive conservation benefits if sold as cheaper and less preferred substitutes, following the same principles outlined for farmed products. Policymakers may be challenged with determining whether such products are marketed as sufficiently distinct; however, this may be worth investigating due to the potential benefits provided.

Consider the example of mammoth ivory, which somewhat resembles elephant ivory but is sufficiently distinct to be recognized by a reasonably trained eye.[7] Since the 1989 CITES ban on elephant ivory trade, Russia started exporting ivory harvested from previously frozen carcass remains of extinct woolly mammoths buried beneath the Arctic tundra. An empirical analysis of trade data found that, consistent with the theory above, introducing the mammoth ivory substitute initially boosted the effectiveness of the 1989 CITES ivory ban for elephant conservation (Farah & Boyce, 2019). A subsequent overall increase in ivory demand in the early 2000s caused a subsequent increase in elephant poaching, but this would likely have been significantly greater in the absence of the competing mammoth ivory supply. The authors found that without competition from mammoth ivory trade, raw elephant ivory prices would have been double their 2019 rates. Contrary to the fears of some conservationists and policymakers, the legal mammoth ivory trade appears to be protecting living wild elephant populations from poaching.

Notes

1 The actual landscape of preferences can be quite complicated and might vary between consumers and from product to product. Not all consumers prefer the same thing—for example, in the case of products that act as food or medicine, some may prefer wild-harvested versions, believing that they are more natural and potent, whereas others might consider farmed products more hygienic and therefore safer. See Hinsley and 't Sas-Rolfes (2020) for a full discussion on this topic.

2 If trade in the wild good is illegal (and trade in the substitute good is legal), then a consumer's valuation of the wild good may decrease if the consumer has some "distaste" for purchasing an illegal good. If this distaste factor is sufficiently large, then it would be possible for $V_w < V_s$ to hold. Throughout this chapter, we will assume that this distaste factor—at least relative to how much lower consumers value the substitute good—is low enough that $V_w \geq V_s$ for all consumers.

3 The economic model can be extended to allow a buyer to purchase more than one unit. Such an extension, however, would make the analysis significantly more complicated without adding any substantive insights.

4 More specifically, the market supply curve for the wild good is the horizontal summation of the marginal cost curves of all wild good sellers. Similarly, the market supply curve for the substitute good is the horizontal summation of the marginal cost curves of all sellers of the substitute good.

5 If we instead assume that the marginal cost curve for the substitute good starts out higher than the marginal cost curve for the wild good at quantity 0, then the price of the wild good must be higher than the price of the substitute good in the non-trivial case in which the substitute good is actually bought and sold in the market.

6 In addition to the previously mentioned possibility of subsidizing substitutes, governments and NGO could engage in information campaigns that highlight the virtues of purchasing and using the substitutes over the wild good.

7 Several features can be used to distinguish the two, including color, structure, and texture, with the most reliable being the angle of the natural grain lines ("Schreger lines"), which are generally less than 90° in cross-section in mammoth ivory and greater than 115° in cross-section in elephant ivory.

References

Farah, N., & Boyce, J. R. (2019). Elephants and mammoths: The effect of an imperfect legal substitute on illegal activity. *Environment and Development Economics, 24*(3), 225–251. https://doi.org/10.1017/S1355770X18000554

Hinsley, A., & 't Sas-Rolfes, M. (2020). Wild assumptions? Questioning simplistic narratives about consumer preferences for wildlife products. *People and Nature, 2*(4), 972–979. https://doi.org/10.1002/pan3.10099

Meeks, D., Morton, O., & Edwards, D. P. (2024). Wildlife farming: Balancing economic and conservation interests in the face of illegal wildlife trade. *People and Nature*, *6*(2), 446–457. https://doi.org/10.1002/pan3.10588

Phelps, J., Carrasco, L. R., & Webb, E. L. (2014). A framework for assessing supply-side wildlife conservation: Wildlife farming and cultivation. *Conservation Biology, 28*(1), 244–257. https://doi.org/10.1111/cobi.12160

Tensen, L. (2016). Under what circumstances can wildlife farming benefit species conservation? *Global Ecology and Conservation, 6*, 286–298. https://doi.org/10.1016/j.gecco.2016.03.007

6 The Economic Model with Substitute Goods and Asymmetric Information

Introduction and Overview

In general, the substitute goods may be so similar—or even bio-identical—to the wild goods that buyers cannot easily distinguish between the two types of goods. Examples of such cases would include many cultivated plants and farmed animal products.[1] In this case, we would not have two separate markets—one for each type of good, as in Chapter 5—coexisting and operating at the same time. This follows since—without the ability to tell the two kinds of goods apart—transactions cannot be dependent on which type of good is being bought or sold. For this scenario, both types of goods would be sold in the same market for the same price, unless there are mechanisms like a credible permit system to allow buyers to differentiate among the goods in the market.

In this chapter, we assume that the sale of wild goods is illegal, and we consider how the market for wildlife goods operates when buyers possess incomplete information about the types of goods being sold, but sellers know exactly what they are selling. Economists refer to a market with such conditions as one with *asymmetric information*. We will examine two cases. In the first case, we assume that sale of the substitute goods is legal, but there are no accompanying policies or regulations requiring labeling or identifications of these goods. Another way to view this case is that, even if regulations and labeling laws exist to enable the legal sale of the substitute good, they are so weak that wild-good sellers can easily pass off their product as the substitute good without incurring any penalties or costs imposed by legal authorities. This assumption may not be realistic because policymakers and government agencies would presumably be reluctant to allow the sale of substitute goods without imposing some stringent regulations on this trade. However, the analysis of this scenario—which is based on the pioneering work of Nobel Prize-winning economist George Akerlof[2]—will serve as a benchmark case that gives us a basis with which to compare other more realistic settings.

For the second case, we will consider one of these alternative settings: specifically, we assume that there exists a permit system and that sellers are allowed to sell the substitute goods only if they acquire a permit. We assume,

DOI: 10.4324/9781003286318-6

however, that—like the buyers—the permit-granting agency cannot perfectly distinguish between the wild goods and the substitute goods. This implies two things in particular: (1) even the wild-good sellers would be able to acquire a permit since permit-issuing officials would not be able to identify the wild goods and (2) following the first point, buyers may still be unsure of what type of good they are buying even when the good is sold with a permit.

In the parlance of economics, the permit could serve as a *market signaling* device (Spence, 1973)—it is a way for substitute-good sellers to "communicate" to the buyers that what they are offering is the substitute good. However, given the ability of the wild-good sellers to also acquire a permit, this signaling mechanism may not be reliable or credible.

We will allow for the possibility that the permit-granting agency—even if it cannot perfectly distinguish between the wild and the substitute good—has some limited ability to tell the difference between the two types of goods. This means that it would be easier for the substitute-good sellers to acquire a permit compared to the wild-good sellers. How this assumption will be operationalized in our economic model will be made clearer and more precise once we present the model.

The Wildlife Market with No Permit System

We use the same modeling framework as the one presented in the previous chapter of this book. However, we simplify the model in a few respects so that we can use a purely graphical approach to analyze the wildlife goods market with asymmetric information. A formal, mathematical presentation of the model and results is given in the technical appendix for this chapter (available at: https://www.frederickchen.org/publications/the-economics-of-wildlife-trade-technical-appendices).

The Modeling Framework

Demand

Let $V_s \geq 0$ and $V_w > 0$, respectively, denote a buyer's willingness to pay (WTP) for the substitute good and the wild good. As before, we assume that each buyer can buy at most one unit of good (whether it be the wild one or the substitute one) and that buyers value the wild good more than the substitute good, i.e., $V_w > V_s$. The buyers can differ from each other in their values of V_w and V_s.

In contrast to the analysis in the previous chapter, because of incomplete information on the part of buyers regarding the type of goods being sold in the market, there is only one price, P, for all wildlife goods. Due to their uncertainty about the goods being sold, the buyers need to make predictions about the likelihood of getting a wild good or a substitute good if they make a purchase, and they make use of these predictions—as well as the prevailing market price—to determine whether to buy a good in the market or not.

We will formalize the buyers' prediction of the chance of getting a wild good or a substitute good by using β, where $0 \le \beta \le 1$, to denote the buyers' estimate of the probability of getting a wild good if they make a purchase. This means then that their estimate of the probability of getting a substitute good is $1-\beta$. For convenience, we will henceforth simply refer to β as the buyers' *belief* regarding the type of good being sold.

Given the buyers' uncertainty and their belief β, we will define their WTP for a good in the market as the expected value of a wildlife good for sale, denoted by EV. This is a weighted average of their WTP for the wild good and their WTP for the substitute good, with β and $1-\beta$ serving as the weights: $EV = \beta V_w + (1-\beta)V_s$. Therefore, given price P and expected value of a good EV, a buyer would choose to make a purchase if $EV - P > 0$; and the buyer would be indifferent between making a purchase or not if $EV - P = 0$. If $EV - P < 0$, then the buyer would definitely choose not to make a purchase.

In line with the economics literature on asymmetric information, we need to impose a consistency restriction on the belief β of buyers to avoid introducing arbitrary or *ad hoc* elements into the model. This restriction follows directly from assuming that the buyers know the sellers' cost of selling (i.e., willingness-to-sell) so that the buyers can make reasonable predictions regarding how the different kinds of sellers would behave given different prices in the market. Following our description below of the supply side of the market, we will provide some intuition for this restriction that the buyers' belief β must satisfy. A more precise, formal presentation of the restriction is given in the technical appendix of this chapter.

Supply

For ease of exposition, we assume without loss of generality that each seller—of the wild good or the substitute good—has only one unit of good available for sale. For a wild-good seller, the cost of selling the good is $c_w \ge 0$; for a substitute-good seller, denote their cost of selling by $c_s \ge 0$. In line with the assumptions that we utilized in previous chapters (and to rule out the trivial case in which the substitute good is not sold in equilibrium), we assume that the cost of selling for the wild-good sellers exceeds the cost of selling for the substitute-good sellers. The cost c_w may vary among the wild-good sellers. Similarly, there may be differences in the value of c_s among the substitute-good sellers. In the figures that follow, $\underline{c}_s$ and $\overline{c}_s$, respectively, denote the lowest and highest values of c_s among the substitute-good sellers. Analogously, $\underline{c}_w$ is the lowest value of c_w among the wild-good sellers. Without loss of generality, we assume that the highest value of c_w among the wild-good sellers is unbounded in the figures.

Sellers know the kind of goods they are selling by assumption, and the derivation of the market supply curve is straightforward: substitute-good sellers would choose to sell if $P > c_s$; similarly, wild-good sellers would choose to sell if $P > c_w$. The supply curve for this market would look as shown in Figure 6.1.

Figure 6.1 The supply curve in the wildlife goods market with asymmetric information.

Market Equilibrium

We first discuss informally the restriction that the buyers' belief β—recall that this is the buyers' estimate of the probability of getting a wild good—should satisfy now that we know the sellers' decision rule regarding supply. In plain language, assuming that buyers know the values of c_s and c_w in the market, or, more precisely, their distribution among the sellers—and that they can anticipate how the sellers would behave at different prices—the value of β at any given price P should be equal to the proportion of wild goods for sale at that price P.

To help fix ideas, suppose the market price P is high enough that it exceeds c_s for all the substitute-good sellers and it exceeds c_w for at least some of the wild-good sellers. If buyers know how different types of sellers decide whether to sell their goods given different prices, then buyers' estimate—in terms of probability—of how likely they are to acquire a wild good in the market should they choose to make a purchase at price P should be equal to the number of wild goods for sale at price P divided by the total number of goods for sale at P.

On the other hand, suppose the market price P is low enough that none of the wild-good sellers choose to sell, but it is just high enough that at least some of the substitute-good sellers choose to sell their products. In this case, if buyers know the cost of selling for the different types of sellers, they could deduce that only the substitute goods would be for sale at price P. This means that their probability of acquiring a wild good at price P is 0.

In an equilibrium of the market, the price P is such that the wildlife good market clears, i.e., $S(P) = D(P)$, where $D(P)$ denotes the demand for goods at price P, and $S(P)$ denotes the supply of goods at price P. Note that $D(P)$ depends on the buyers' belief β, which—as stated above—is itself dependent on the supply of the wild and the substitute goods at price P.

Properties of Market Equilibrium

Graphically, the market equilibrium is the point at which the supply and demand curves in the wildlife goods market intersect. Before we discuss the main properties of the market equilibria, it is instructive to consider in more detail the shape of the demand curve in this market, since it critically affects how the market behaves.

Unlike the market demand curve that we saw in Chapter 4, which is always downward sloping, the demand curve in our market with asymmetric information could have an upward-sloping portion—though it need not have one. The intuition for this is surprisingly simple. If there is asymmetric information in the market, a change in price could give rise to two opposing effects on consumer demand. On the one hand, when price increases, a consumer's net benefit of purchasing a good—all else equal—decreases; this *price effect* tends to reduce the quantity demanded with a rise in price. This is the reason why most demand curves we come across are downward sloping.

On the other hand, when *P* is high enough that some—but not all—wild-good sellers choose to sell, then increasing the price would induce more wild-good sellers to sell. This means the fraction of goods for sale that are wild would rise as *P* becomes higher—thereby increasing the probability that a buyer would get a wild good by making a purchase—which would cause the buyers' WTP in the market to go up. This effect on its own—which we could call the *quality effect*—in turn would tend to raise the quantity demanded as price increases.

With asymmetric information in the market, therefore, whether a rise in price would ultimately increase or reduce demand depends on the relative magnitudes of the price and quality effects: if the price effect dominates, then the demand would go down; if the quality effect dominates, then the demand would go up instead. Note that in the model of Chapter 4—where there is no asymmetric information because there is only one type of good—the quality effect is absent; hence, an increase in price in that context unambiguously reduces demand.

Another property to note about the demand curve when there is asymmetric information is that—despite the quality effect—it cannot be upward-sloping all the way. Take, for instance, the portion of the demand curve when the price *P* is low enough that only the substitute-good sellers choose to sell. For this segment, any small increase in *P* would only bring about a price effect—there cannot be a quality effect—and the curve must have a negative slope.

At the other extreme, if the price is so high that all the substitute-good and wild-good sellers would choose to sell, then no increase in price beyond this point would result in any quality effect since the fraction of goods for sale that are wild cannot increase any further. This means that, for prices in this region, the demand curve must be downward sloping. Figure 6.2 shows what the demand curve in a market with asymmetric information could look like.

Now that we have an idea of the general shape that the demand curve could have, we highlight three major properties of the market for wildlife goods when there is asymmetric information. While the exposition below employs a purely graphical approach, we note that all the results presented in the text could also be derived mathematically using numerical examples, which we do in the technical appendix for this chapter.

Adverse Selection

First, the presence of the substitute goods in the market may reduce buyers' WTP—and, hence, the equilibrium market price of wildlife goods—so much that the wild-good sellers all exit the market, leaving behind only the substitute-good sellers in the market. In economics, such a phenomenon—when the introduction of low-quality goods drives out high-quality goods which are more expensive to produce—is known as *adverse selection*. This result is shown in Figure 6.3.

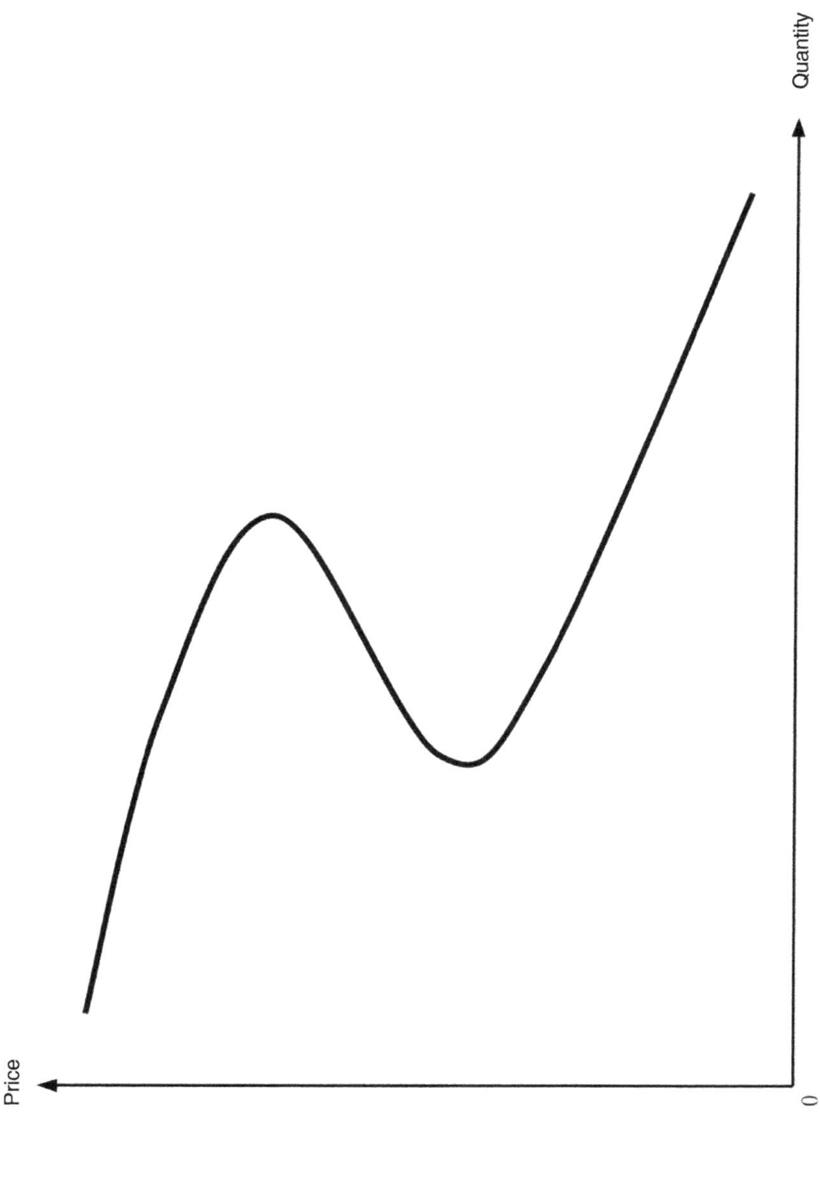

Figure 6.2 The market demand curve in the wildlife goods market with asymmetric information could have an upward-sloping portion due to the quality effect.

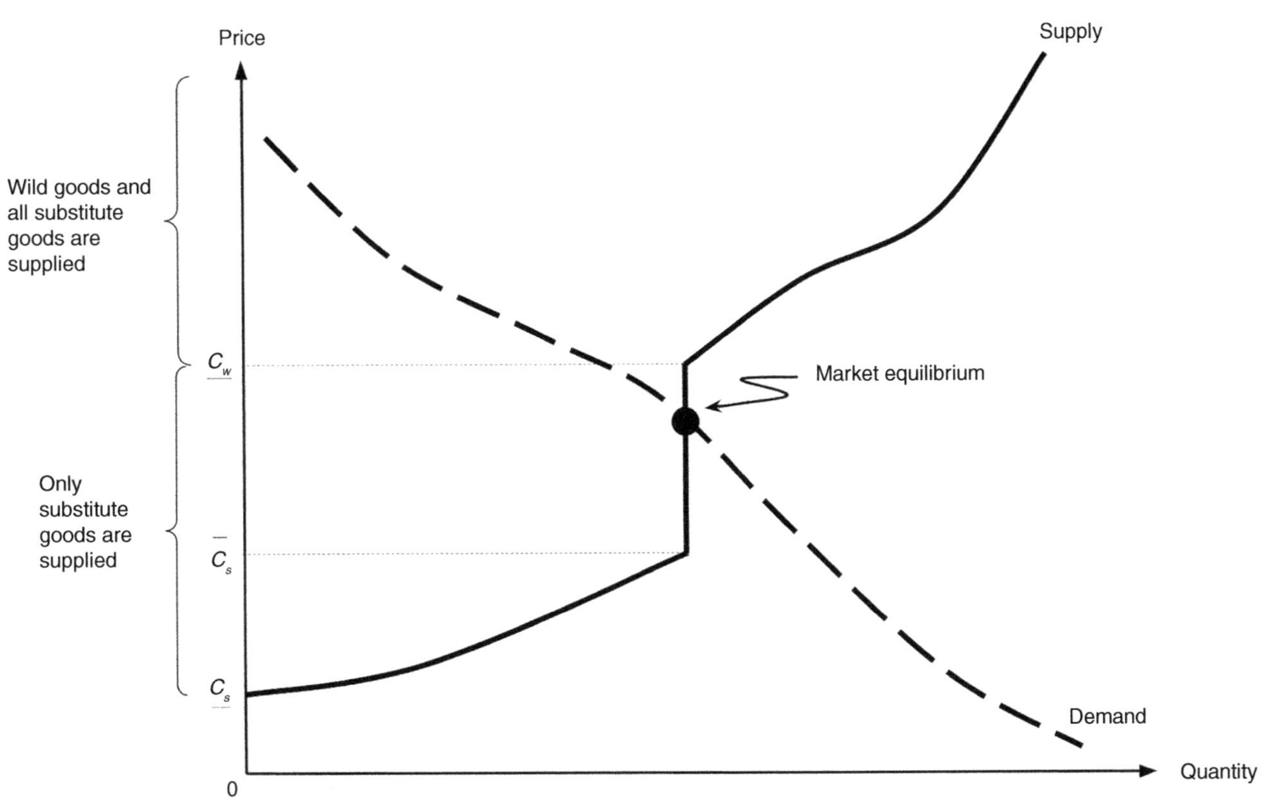

Figure 6.3 Market equilibrium with adverse selection. Because the equilibrium price is below the minimum price at which wild-good sellers are willing to sell, only the substitute goods are sold in equilibrium.

Note that adverse selection need not necessarily emerge in the market with asymmetric information. For it to occur, the buyers' valuation of the substitute good must be sufficiently low or the ratio of substitute-good sellers to wild-good sellers must be sufficiently high. This implies that if the substitute good is considered significantly inferior to the wild good or if there are enough substitute goods on the market, then all trade in wild goods would cease.

Multiple Equilibria

The second major point we highlight about the wildlife market with the substitute good and asymmetric information is that, unlike the case without asymmetric information, there can be more than one market equilibrium. As Figure 6.4 shows, when the demand curve has an upward-sloping portion, it is possible—though not necessarily so—for the demand curve and the supply curve to intersect at three points. Note that if the demand curve is always downward sloping, then there exists only one market equilibrium. This implies that the quality effect we discussed previously must be sufficiently strong for multiple equilibria to arise.

Intuitively, multiple equilibria can coexist because the presence of asymmetric information can give rise to a positive feedback loop in the market. To see this, suppose the buyers somehow expect many wild goods to be sold in the market. Note that this belief implicitly implies that many substitute goods would be sold as well since any price that induces wild-good sellers to sell would be high enough to induce the substitute-good sellers to sell. The expectation that there are many wild goods being sold leads the buyers to be willing to pay a relatively high price for a wildlife good. If this price they are willing to pay exceeds the wild-good sellers' cost of selling, then the buyers' expectation—that there would be many wild-good sellers offering their product for sale—would be consistent with what the sellers actually choose to do, and we would have an equilibrium in which the wild good and the substitute good are both sold in the market. Note in this instance that there would be no adverse selection in equilibrium.

On the other hand, suppose the buyers have the opposite—one could say, pessimistic—expectation and believe that no wild goods are being sold in the market. This expectation leads them to be unwilling to pay a high price for a wildlife good. If the maximum price they are willing to pay to buy a wildlife good falls short of the wild-good sellers' cost of selling, then there would indeed be no wild goods in the market. In this case, again, the buyers' expectation of what the wild-good sellers would do is consistent with what they actually choose to do, and we would arrive at a market equilibrium in which no wild goods are sold.

What the above discussion shows is that, with asymmetric information in the market, buyers' expectation of what kind of goods they expect to be sold can be self-fulfilling: if they expect the average quality of goods in the

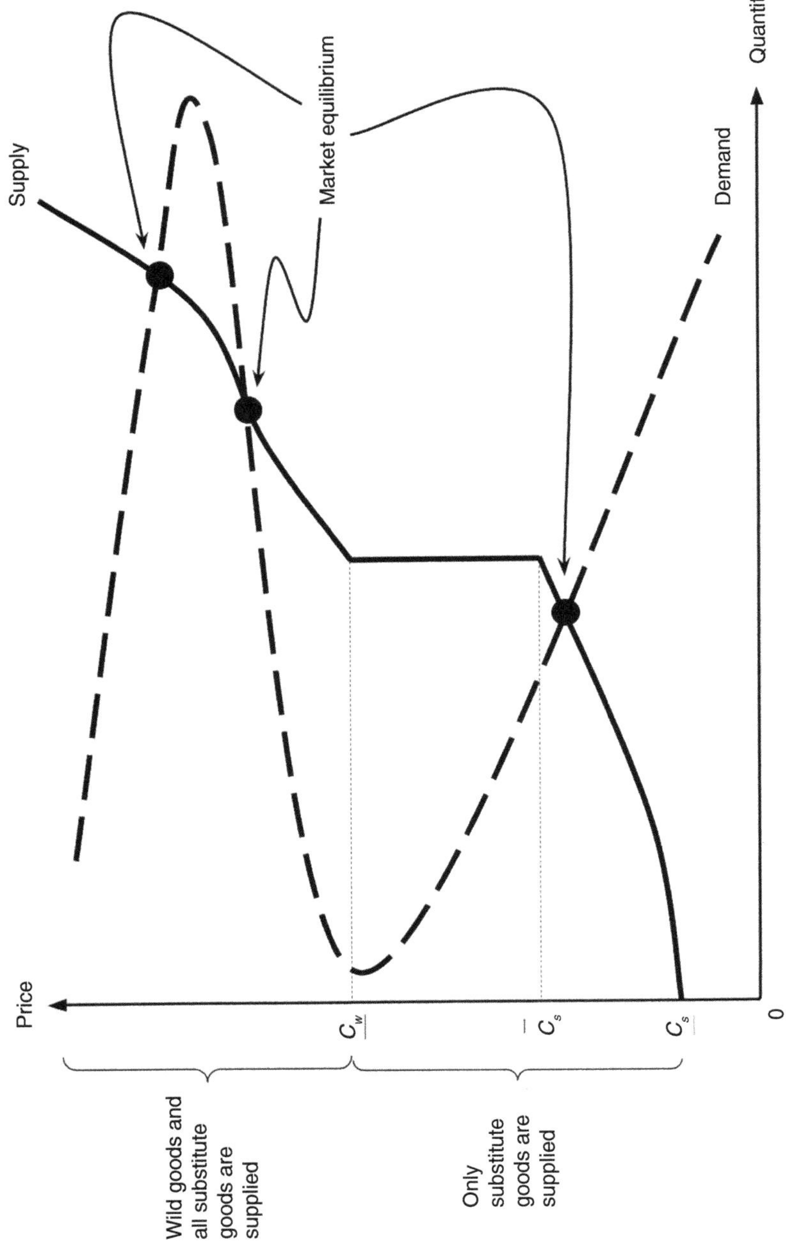

Figure 6.4 The possibility of multiple market equilibria existing due to the demand curve having an upward-sloping portion. Note that the higher the equilibrium price, the greater the amount of wild good supplied in equilibrium.

market to be high—where the wild good is considered the high-quality good and the substitute good is the low-quality good—and behave accordingly, then the average quality of goods sold has the tendency to be high; on the other hand, if the buyers expect the average quality of goods to be low and they behave according to this belief, then the average quality of goods would trend downwards. It is this positive feedback mechanism created by buyers' expectations and their accompanying actions that leads to the possibility of multiple market equilibria.

Reduction of Wild Goods Even without Adverse Selection

The third point we will highlight about the wildlife good market with asymmetric information is that, even when adverse selection does not occur in equilibrium, the presence of the substitute good could decrease the quantity of the wild good supplied—i.e., relative to the level in the absence of the substitute-good sector. The intuition for this is straightforward. As discussed earlier, when the substitute good is available but buyers are not able to distinguish between the wild good and the substitute good, their WTP for a wildlife good—and, hence, the market price of wildlife goods—will decrease. If the wild-good sellers differ in their cost of supply, then the decline in price—even if not large enough to result in adverse selection—would cause the higher-cost sellers to drop out of the market, thereby reducing the amount of wild goods being sold. Figure 6.5 shows this result graphically.

Policy Implications

Effects of Subsidizing the Substitute-Good Sector

Let us now consider the effects of a subsidy for the sellers of the substitute good. To avoid trivial cases, we assume here that there is no adverse selection in equilibrium when the substitute-good sector is not subsidized. Intuitively, a subsidy for the substitute-good sector, by raising the supply of the substitute good at every price of the wildlife good, would lower buyers' expected value—and, hence, their WTP—for a wildlife good. This would, in turn, put downward pressure on the market price and lower the incentives of the wild-good sellers to supply their good. Therefore, we should expect the supply of the wild good in equilibrium to decrease as a result of subsidizing the substitute-good sellers.

Note, however, that because multiple market equilibria can obtain in the market with asymmetric information, it is not entirely straightforward to compare the wildlife good market without a subsidy for the substitute-good sellers to one in which they are subsidized for their production. This follows simply because, when multiple equilibria coexist, a comparison of the two markets will in general depend critically on which equilibrium we select for the comparison exercise.

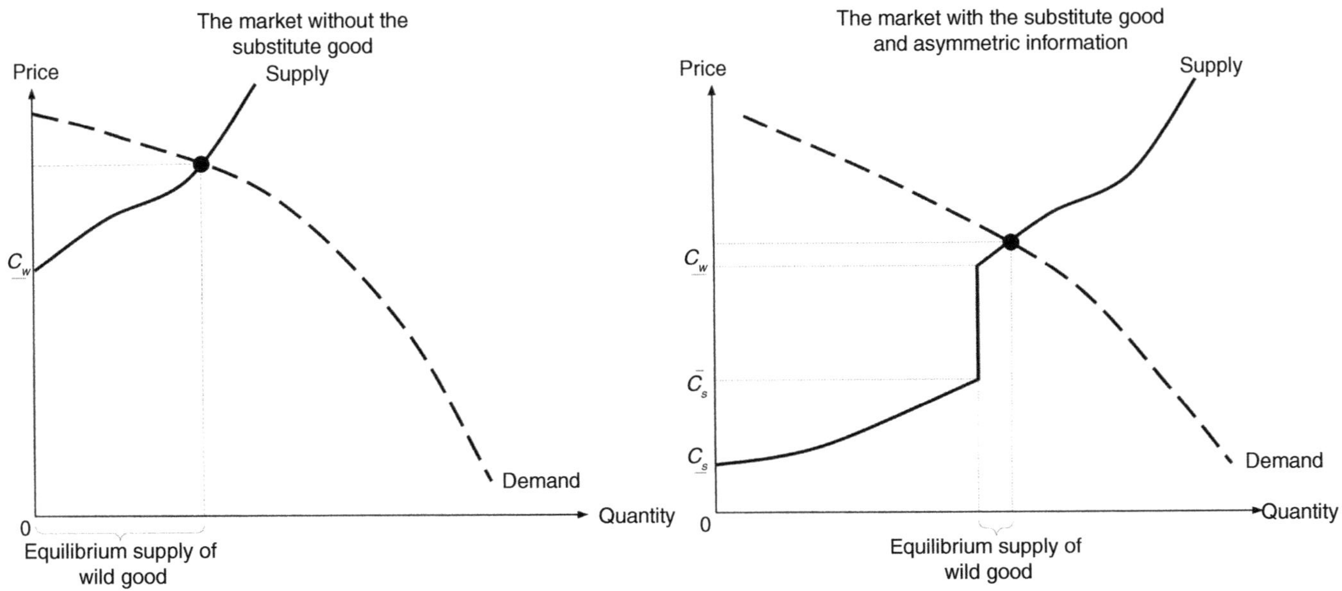

Figure 6.5 The presence of the substitute good lowers the equilibrium supply of the wild good even in the absence of adverse selection. When the substitute good is available in the market and there is asymmetric information, the demand for wildlife good is lower at every price compared to the demand in the market without the substitute good. This occurs because the presence of the substitute good lowers buyers' WTP for a wildlife good. This, in turn, lowers the equilibrium price in the market with the substitute good, which reduces the equilibrium supply of the wild good.

To make our analysis of the effects of a subsidy for the substitute-good sellers more concrete, we will henceforth concentrate on the equilibrium with the highest price when multiple equilibria obtain. Figure 6.6 shows that, with such a restriction in place, the intuition described above is correct: subsidizing the substitute-good sector lowers the equilibrium supply of the wild good.

Effects of Stronger Anti-Poaching Measures

In our framework, policies aimed at curbing poaching—for example, tougher penalties for poaching, stricter enforcement of anti-poaching laws, or better physical protection of wildlife—could be thought of as raising the cost of supplying the wild good, c_w. Such policies—by lowering the availability of the wild good at every price—should, intuitively, reduce the equilibrium supply of the wild good.

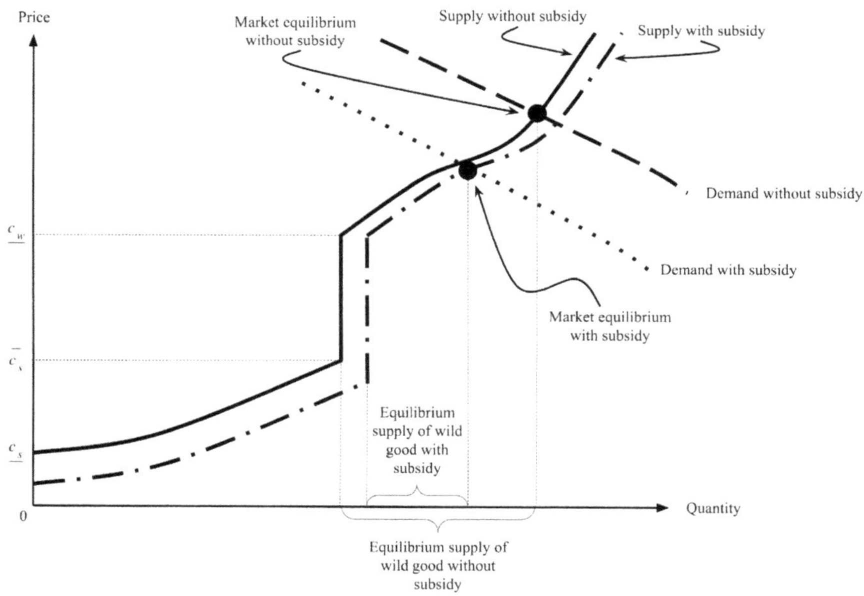

Figure 6.6 A subsidy for the substitute-good sellers lowers the equilibrium supply of the wild good. The subsidy shifts the supply curve down and out. Because the subsidy increases the supply of the substitute good at every price, the buyers' WTP for a wildlife good—which is based on the proportion of goods for sale that are wild—decreases, which means there is lower demand at every price above $\underline{c_w}$. The figure shows only a portion of the demand curves. Note that if multiple equilibria coexist, then the equilibrium with the highest price must occur at a downward-sloping segment of the demand curve.

As noted in the above discussion regarding the impact of subsidizing the substitute-good sector, a comparison of market equilibria before and after the implementation of a policy is complicated by the fact that multiple equilibria can coexist when there is asymmetric information. Figure 6.7 shows, however, that if we confine attention to the equilibrium with the highest price when there exist multiple equilibria, then it is indeed the case that increasing the cost of selling the wild good leads to a reduction in the equilibrium supply of the wild good. Moreover, more effective anti-poaching measures may make adverse selection more likely by eliminating the possibility of having market equilibria in which the wild good is sold.

Effects of Demand-Side Promotion of the Substitute Good

We now turn our attention to examining the effects of promotional or marketing campaigns aimed at stimulating demand for the substitute good. As in the previous chapter, such efforts targeting the demand side could be modeled as increasing the buyers' WTP for the substitute good.

As in the market with complete information discussed in Chapter 5, policies designed to enhance the value or attractiveness of the substitute good for buyers could in fact ***increase*** the quantity supplied of the wild good in the market with asymmetric information. This happens because a higher value of V_s—which denotes buyers' WTP for the substitute good—raises their WTP for a wildlife good—all else equal—when they cannot distinguish between the wild good and the substitute good. This results in a higher market price. In turn, the higher price induces more wild-good sellers to enter the market and sell their products, as shown in Figure 6.8. Here, again, we focus on the equilibrium with the highest price if multiple equilibria coexist in the market.

The Wildlife Market with a Permit System

In the previous section, we considered a wildlife good market with asymmetric information in which the substitute-good sellers do not have the option to obtain special approvals or allowances to sell their products. While this may not be a realistic assumption, the analysis provides us with a benchmark case with which we can compare alternative scenarios.

In this section, we will consider the same wildlife good market with asymmetric information; however, we assume here that a permit system is in place whereby sellers must acquire a permit at some cost if they wish to sell their goods in the market legally. The analysis, therefore, allows us to see how greater governance of the market through, for example, the implementation of a permit system, can affect the supply of the wild good in equilibrium.

We note that a rigorous analysis of a permit system for selling synthetic wildlife goods was already presented in Chen and 't Sas-Rolfes (2021). In that model, however, it is assumed—unlike here—that the wild good and the substitute synthetic good are perfect substitutes, i.e., buyers' WTP is the

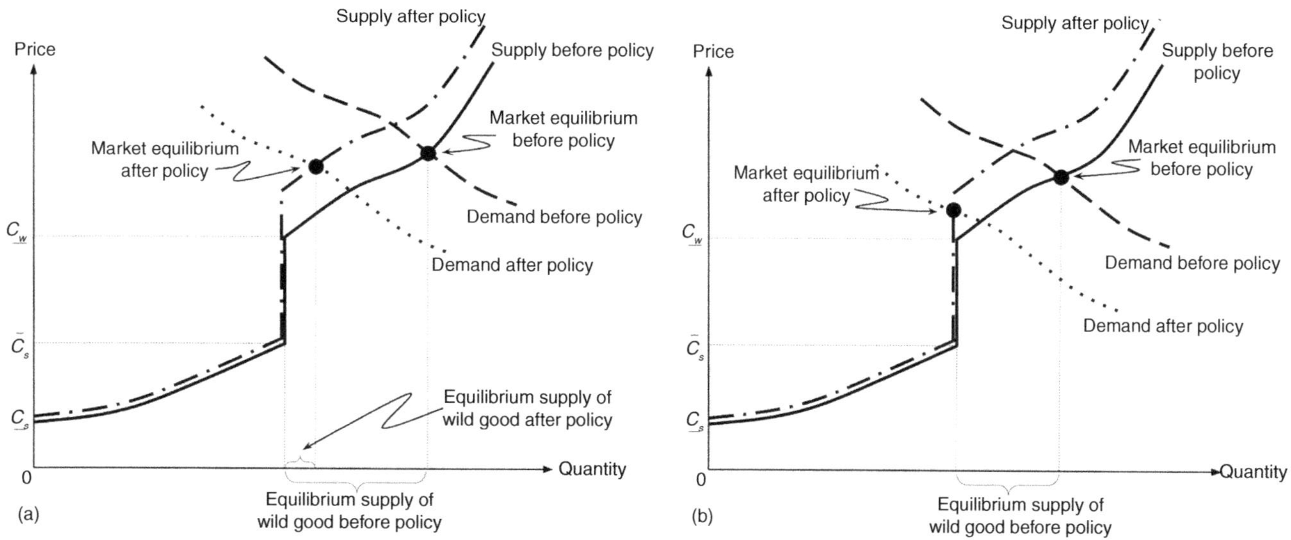

Figure 6.7 Implementing a policy with more effective anti-poaching measures reduces the equilibrium supply of the wild good. Because the policy reduces the supply of the wild good at every price above $\underline{C}_w$, the buyers' WTP for a wildlife good—which is based on the proportion of goods for sale that are wild—also decreases; this lowers the demand at every price above $\underline{C}_w$. (a) The policy does not lower demand sufficiently to cause adverse selection. (b) The reduction in demand resulting from the policy is large enough to cause adverse selection.

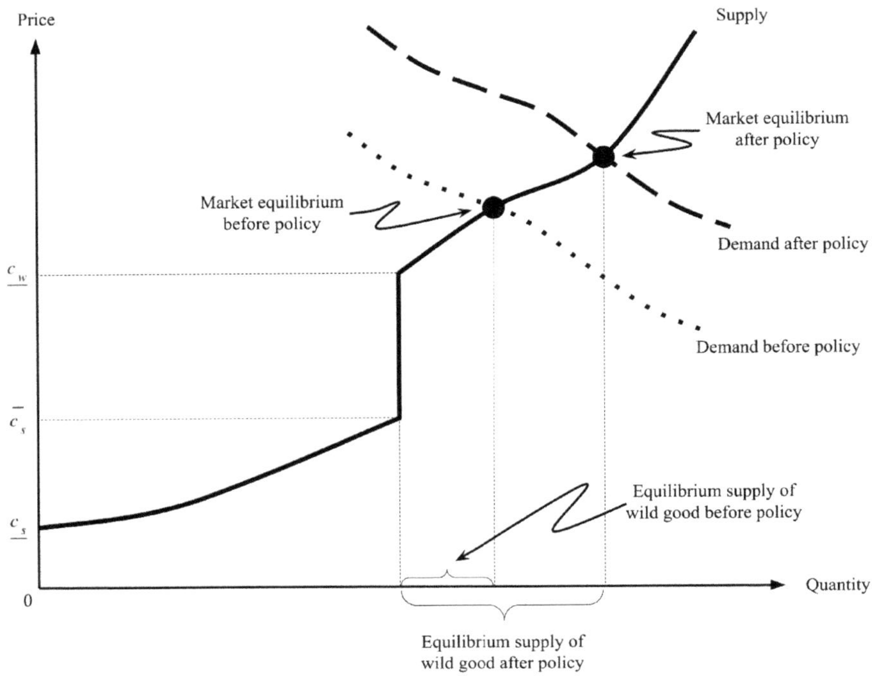

Figure 6.8 A policy to promote the substitute good increases the equilibrium quantity supplied of the wild good. The policy increases buyers' WTP for the substitute good, which in turn raises demand in the wildlife good market at every price. This leads to an increase in the equilibrium price and the equilibrium quantity supplied of the wild good.

same for the two types of goods. Hence, our analysis in this section, which assumes that the buyers' WTP for the two types of goods are different—can be considered as a complement to that earlier exposition.

A Model of a Permit System

Let us assume that the following permit system is implemented. A seller must obtain a permit to sell their product legally. While buyers cannot ascertain for themselves whether a wildlife good for sale is the wild one or the substitute good, we assume that the buyers can observe whether a seller in the market has a permit or not. A permit then could potentially serve as a market signal of what type of wildlife good is being sold.

A permit is costly for a seller to obtain: for example, the application process may be time-intensive and require a not-so-insignificant amount of effort to complete; or financial expenses must be incurred to obtain one. It is certainly possible that getting a permit requires both a financial outlay and expenditures of time and effort. To simplify the analysis, we summarize and aggregate all the spending—both pecuniary and nonpecuniary—that a seller must incur to get a permit by a single value $\gamma > 0$, which we refer to as the *cost of a permit*.

The spirit—and intent—of the permit system is to allow for a legal trade of the substitute good (and cut down on illegal trade of the wild good). We assume, however, that—like buyers in the market—the authorities that issue the permits also cannot distinguish between the wild good and the substitute good. This means that both the substitute-good sellers and the wild-good sellers could obtain permits to sell their wares. However, it may require additional time, effort, or expenses for the wild-good sellers to "disguise" themselves as the substitute-good sellers for the purposes of applying for a permit. Hence, the cost of a permit γ could differ by seller-type, with it being higher for the wild-good sellers. We will thus assume that $\gamma = \gamma_w$ for a wild-good seller and that $\gamma = \gamma_s$ for a substitute-good seller, where $\gamma_w > \gamma_s$.

Because buyers can see whether a good is sold with a permit or not, there are two submarkets for wildlife goods, each with its own price: one for goods sold with a permit and one for goods sold without a permit. Let P_Y denote the price of a wildlife good that is sold with a permit and let P_N denote the price of a wildlife good that is sold without a permit. Below, we describe the buyers' and sellers' decision problems in these markets more precisely.

Supply

For a substitute-good seller, the net gain or profit from obtaining a permit and selling their good legally is $P_Y - c_s - \gamma_s$. If the seller does not obtain a permit and sells the good illegally, then the seller may get caught and penalized by law enforcement agencies for not having a permit. Let f denote the (expected) cost to a seller from selling goods without a permit[3]. The profit to a substitute-good seller from selling their good without a permit is thus $P_N - c_s - f$. Therefore, it is better to get a permit if $P_Y - c_s - \gamma_s > P_N - c_s - f$, or, after rearrangement, $P_Y - P_N > \gamma_s - f$; it is better not to get a permit if $P_Y - P_N < \gamma_s - f$; and the two options are equally good if $P_Y - P_N = \gamma_s - f$. Note that substitute-good sellers would choose to supply their good only if the profit from their best option is nonnegative. The term $\gamma_s - f$ could be interpreted as the relative or net—as opposed to absolute—cost to a substitute-good seller of obtaining a permit. While getting the permit itself requires a cost of γ_s, having the permit eliminates the cost f. Hence, on net, the seller is incurring the cost $\gamma_s - f$ from getting a permit.

Analogously, for a wild-good seller, the profit from obtaining a permit and selling is $P_Y - c_w - \gamma_w$, while the profit from selling without a permit is $P_N - c_w - f$, where we are assuming that the cost of getting caught without a permit is the same for both types of sellers. Therefore, it is better to get a permit if $P_Y - P_N > \gamma_w - f$; it is better not to get a permit if $P_Y - P_N < \gamma_w - f$;

and the two options are equally good if $P_Y - P_N = \gamma_w - f$. Note that wild-good sellers would also choose to supply their good only if the profit from their best option is nonnegative. Analogous to the case of the substitute-good sellers, the relevant cost that a wild-good seller takes account of when deciding whether to obtain a permit or not is the net cost of a permit $\gamma_w - f$.

Let us define the following notations that will be used in the analysis later. Let $S_{wY}(P_Y, P_N)$ and $S_{wN}(P_Y, P_N)$, respectively, denote the supply of the wild good with and without a permit given prices P_Y and P_N. Similarly, let $S_{sY}(P_Y, P_N)$ and $S_{sN}(P_Y, P_N)$, respectively, denote the supply of the substitute good with and without a permit given prices P_Y and P_N. The total supply of goods sold with a permit is $S_Y(P_Y, P_N) = S_{wY}(P_Y, P_N) + S_{sY}(P_Y, P_N)$, and the total supply of goods sold without a permit is $S_N(P_Y, P_N) = S_{wN}(P_Y, P_N) + S_{sN}(P_Y, P_N)$.

Demand

Given that wildlife goods can be sold with or without a permit, buyers need to consider the expected value of a good sold with and without a permit. They then need to compare them to the prices P_Y and P_N, respectively, in order to determine, first, whether it is better to buy a good with or without a permit, and second, whether it is worthwhile to buy anything at all at the given prices.

Let EV_Y and EV_N, respectively, denote the expected value of a wildlife good sold with and without a permit. As before, the expected value of a wildlife good is a weighted average of the buyers' WTP for the wild good and their WTP for the substitute good, where the weights are given by the respective probabilities of obtaining the different types of goods in the market. An additional factor to consider for buyers when a permit system exists is the extent to which buying a good without a permit diminishes a buyer's WTP for the good due to, for example, the stigma of or moral discomfort from purchasing a good that is sold illegally. Let us denote by m the amount by which a buyer's WTP for a good decreases when it is purchased without a permit. For convenience, we will henceforth refer to m as the *moral cost* of purchasing a good that is sold illegally.

A buyer's WTP for a good sold with a permit is simply its expected value: $EV_Y = \beta_Y V_w + (1 - \beta_Y) V_s$, where β_Y denotes the probability that a good sold with a permit is wild. A buyer's WTP for a good sold without a permit is its expected value, $EV_N = \beta_N V_w + (1 - \beta_N) V_s$, where β_N is the probability that a good sold without a permit is wild, minus the moral cost m, i.e., this WTP is $\beta_N V_w + (1 - \beta_N) V_s - m$. (Note that $1 - \beta_Y$ and $1 - \beta_N$, respectively, denote the probabilities that a good sold with and without a permit is a substitute.)

As in the model without the permit system, the probabilities β_Y and β_N must be consistent with what the sellers actually do. Intuitively, β_Y and β_N, respectively, should be equal to the proportions of wild goods sold in the market with and without a permit. A more formal definition of β_Y and β_N is given in the technical appendix for this chapter.

The net benefit of buying a good sold with a permit is $EV_Y - P_Y$, and the net benefit of buying one sold without a permit is $EV_N - m - P_N$. A buyer would thus buy the option with the higher net benefit provided that its net benefit is nonnegative. (If the two options yield the same net benefit, then a buyer would go with either one.) This rule allows us to derive the demand curves in our market. Let $D_Y(P_Y, P_N)$ and $D_N(P_Y, P_N)$, respectively, denote the demand for goods sold with and without a permit given prices P_Y and P_N. We note that both the demand for goods sold with a permit and the demand for goods sold without a permit depend on both prices P_Y and P_N. This is the case because, first, the decision of whether to purchase a good sold with or without a permit depends on both prices; and, second, the decision of sellers to sell a good with or without a permit depends on both prices, and the sellers' decisions impact the probabilities β_Y and β_N. These, in turn, affect the demand for goods sold with and without a permit.

Market Equilibrium

Since there are two submarkets operating when there is a permit system in place—one for goods sold with a permit and one for goods sold without one—the market equilibrium condition is that both markets must clear. More precisely, in equilibrium, the prices P_Y and P_N must be such that $D_Y(P_Y, P_N) = S_Y(P_Y, P_N)$ and $D_N(P_Y, P_N) = S_N(P_Y, P_N)$. We note that, because there are two separate prices that need to be determined simultaneously in equilibrium—rather than just one as in the market without a permit system—it is difficult to analyze the set of market equilibria of this model using a purely graphical approach as this entails visualizing three-dimensional figures. In what follows, we will simply present the main results of the analysis as well as intuitive explanations for them, and we will relegate the formal mathematical analysis to the technical appendix for this chapter.

We note that if the cost of obtaining a permit is too high, i.e., $0 < P_Y - P_N < \gamma_s - f$ so that the gain from selling a good with a permit over the gain from selling one without a permit is less than the net cost of a permit to a substitute-good seller, then no seller would ever acquire a permit in equilibrium. This would lead us to the same outcome(s) as the model without a permit system in place. To examine whether having a permit system in place can affect the wildlife goods market in a substantive manner, we will henceforth focus on the nontrivial scenario in which $P_Y - P_N > \gamma_s - f$, i.e., substitute-good sellers would prefer to obtain a permit over selling their goods without one.

Properties of Market Equilibrium

There are two cases to consider.

Separating Equilibrium

For this case, which requires $\gamma_s - f < P_Y - P_N < \gamma_w - f$, only the substitute-good sellers would choose to obtain a permit since a permit would be cost-prohibitive for a wild-good seller to acquire. In such an equilibrium—if it exists—the permit would serve as a perfect signal of a seller's type: any seller offering their product with a permit must be a substitute-good seller, and any seller offering their good without a permit must be a wild-good seller. Such an outcome would deliver the intended result of the permit system in terms of allowing the market to distinguish between the wild goods and the substitute goods.

As shown in the appendix, such a so-called *separating equilibrium* (in the parlance of economists) exists in our market if the parameter values of the model satisfy certain conditions. Intuitively, these conditions say that, first, γ_s must be sufficiently small; otherwise, the substitute-good sellers would have no incentive to sell their products with a permit at all.

A second condition for the existence of a separating equilibrium is that the sum $m+f$ cannot be too high. Recall that m is the reduction in buyers' WTP for a good if it is sold without a permit, and f is the added cost to a seller of supplying a good without a permit when a permit system exists. If m is high, then there would be low demand for goods sold without a permit, driving down their price. If f is large, then the total cost of selling a good without a permit would be high. Both factors would increase wild-good sellers' incentive to obtain a permit, which would make it difficult for the wildlife goods market to reach a separating equilibrium.

Pooling Equilibrium

In this type of equilibrium—which requires $\gamma_w - f \leq P_Y - P_N$—the cost of obtaining a permit is low enough that all sellers—both wild-good and substitute-good ones—would acquire a permit to sell their products. The permit would thus not serve as a reliable signal of whether a good is a wild good or a substitute one. We show in the technical appendix that such a *pooling equilibrium* exists under certain conditions on the parameter values of the model.

At an intuitive level, a pooling equilibrium exists when γ_w is sufficiently low because the wild-good sellers have an incentive to "imitate" the substitute-good sellers by acquiring a permit to sell their goods. In this way, a wild-good seller can mimic a substitute-good seller and try to pass off their wild good as a substitute one in order to get the price P_Y for their good and avoid the penalty cost f that comes from selling a good without a permit.

Additionally, for a pooling equilibrium to exist, the sum $m+f$ must be sufficiently large. A high value of f reduces the attractiveness of selling without a permit. At the same time, a high value of m decreases the price that a good sold without a permit can obtain in the market, further diminishing the value of selling without a permit. Hence, when m or f or both are large, wild-good sellers have more incentive to obtain a permit. The outcome of this, of course, is that in the end all goods come with a permit so that the buyers are no wiser regarding what they are getting compared to the situation in which there is no permit system.

This brings us to the following critical question: Given our result above that there is a pooling equilibrium—in which acquiring a permit conveys no additional information to buyers—when the model parameter values satisfy certain conditions, is there any conservation benefit to implementing such a permit system at all when the market reaches a pooling equilibrium?

Intuitively, one might surmise that, since buyers cannot rely on the permits to distinguish between wild- and substitute-good sellers in a pooling equilibrium, there is no conservation benefit to implementing a permit system if the resulting market equilibrium is a pooling equilibrium. This, however, is not necessarily the case.

To see this, consider a wildlife good market without a permit system in which there is a market equilibrium such that both wild- and substitute-good sellers offer their goods for sale (i.e., an equilibrium without adverse selection). Now, suppose a permit system is introduced, and the parameter values are such that a pooling equilibrium exists, i.e., both wild- and substitute-good sellers sell their goods with a permit in equilibrium. *The supply of wild goods in the pooling equilibrium with a permit system can be lower than the supply of wild goods in a market equilibrium without a permit system*. In other words, even if the permit system does not allow buyers to reliably distinguish between the two types of goods, it can still be useful from a conservation standpoint by decreasing the equilibrium amount of wild good sold. This result follows simply from the fact that, when there is a permit system and the market is in a pooling equilibrium, there is an added cost to wild-good sellers from selling their products—i.e., the cost of obtaining a permit—that does not exist in the market without a permit system in place. This added cost of supply in a market with a permit system thus reduces the incentive of wild-good sellers to sell their goods. (A numerical example demonstrating this result formally is given in the appendix.)

Coexistence of Separating and Pooling Equilibria

We note that, depending on the parameter values of the model, it is possible for the wildlife goods market to have both a separating equilibrium and a pooling equilibrium (an example is given in the appendix). This means that, when both types of equilibrium coexist, the model cannot tell us which equilibrium outcome the market will reach, only that the market should end up in one of the possible equilibria. In terms of policy implications, this result implies that there is no guarantee that implementing a permit system would allow the buyers to distinguish between the wild-good sellers and the substitute-good sellers by observing which sellers have a permit when both types of equilibrium are possible. To ensure that putting in place a permit system would allow the market to differentiate between the wild goods and the substitute goods, our analysis earlier suggests that additional policies that change the values of m, f, γ_s, or γ_w may be needed to complement the permit system.

Using a Permit System with a Separating Equilibrium for Conservation Purposes

Consider a market for a wild good that is responsible for the population decline of some wildlife species. Assume now that a close substitute good becomes available commercially, but that consumers are not able to distinguish between the two types of goods. Would allowing a legal trade in the substitute good and implementing a permit system as described above that leads to a separating equilibrium be helpful from a conservation perspective?

Given the analyses and results derived earlier in this chapter—particularly the results concerning adverse selection—there is a simple and short answer to this question, and that is: *no*, because there is a better policy—and that is to introduce the substitute good into the market without any permit system, thereby creating asymmetric information in the wildlife goods market. As we have seen earlier in this chapter, if the substitute good is valued less by buyers than the wild good, then the availability of the substitute good—without buyers having the ability to distinguish it from the wild good—would reduce demand in the wildlife goods market. If this reduction in buyers' WTP is large enough, adverse selection would result whereby the wild good sellers are driven out of the market entirely. Even if the reduction in demand is not large enough to result in adverse selection, the concomitant decline in the price of wildlife goods in the market would drive down the supply of the wild good (relative to the level prior to the introduction of the substitute good into the market).

Needless to say—but we will say it anyway—any consideration of policy recommendation must balance its intended objectives, feasibility, financial costs of implementation and maintenance, ethics, potential effects on the overall population (not just the target population for the policy), and a host of other factors that policymakers need to account for. When informational problems exist in wildlife goods markets, whether implementing a permit system intended to allow buyers to distinguish between different types of goods is an "optimal" or beneficial policy therefore depends on the calculus of policymaking and the trade-offs that policymakers are willing to make. The economic framework presented above provides us with a better perspective on what these trade-offs are.

Notes

1 There are also instances where one (more common) species may be used as a substitute for another. For example, lion bones have been used as a substitute for tiger bones in the preparation of traditional Asian medicines. The two products are very similar and only highly specialised experts or people able to conduct DNA tests can tell them apart, so lion bone products are easily fraudulently marketed as being tiger bone products.

2 See Akerlof (1970).

3 Note that, as discussed earlier in Chapter 3, the value of *f* is itself the product of two underlying values, namely, the value of the penalty for breaking the law and the expected probability of incurring that penalty.

References

Akerlof, G. (1970). The market for lemons: Quality uncertainty and the market mechanism. *The Quarterly Journal of Economics, 84*(3), 488–500. https://doi.org/10.2307/1879431

Chen, F., & 't Sas-Rolfes, M. (2021). Theoretical analysis of a simple permit system for selling synthetic wildlife goods. *Ecological Economics, 180,* 106873. https://doi.org/10.1016/j.ecolecon.2020.106873

Spence, M. (1973). Job market signaling. *The Quarterly Journal of Economics, 87*(3), 355–374. https://doi.org/10.1016/j.ecolecon.2020.106873

7 The Economic Model with Speculators and Stockpiling

Introduction and Overview

So far, our analysis has ignored the factor of time. In the basic model without intermediaries, time appears irrelevant because there is only one transaction. However, when intermediaries become involved, transactions along the trade chain take place at different times. Most wildlife goods are at least somewhat perishable—wild meat, for example, has a shelf life—but some, for example, elephant ivory, can last for many years and even up to multiple centuries. Demand for wildlife goods may also vary over time, influenced by fashion trends and other factors. For this reason, some intermediaries may decide not to sell the goods they have bought immediately, but rather to store them in the hope that they might sell them at a future date at a higher price than they could achieve in the present. This kind of storage (called "stockpiling") is speculative because the future is uncertain, and demand can either increase or decrease relative to supply. Furthermore, it incurs costs—both the physical costs of storing goods and the "opportunity cost" of keeping one's money invested in such goods rather than in some other income-generating asset. In this chapter we extend our basic model to improve our understanding of how deliberate speculative stockpiling by intermediaries (i.e., "speculators") might affect markets for wildlife goods and, ultimately, conservation.

The Wildlife Market with Competitive Speculators

Because speculation and stockpiling involve inter-temporal choices or decision-making—how many units of wildlife goods to buy now; how many of these units to sell in the future—any model of such behavior must be dynamic, i.e., involve multiple time periods. To keep our analysis as simple and tractable as possible, we will assume that there are only two time periods (period 1 and period 2) in our model. Many of the key insights from a more general multi-period model can be obtained from the two-period version of the model.

There are three types of actors in our model: the consumers of a wildlife good, the producers of the wildlife good, and the speculators. We assume

DOI: 10.4324/9781003286318-7

that no substitute good exists or is available in the market. Once consumers purchase the wildlife goods, they consume or keep those goods so that those goods do not get resold. Speculators, on the other hand, do not consume the wildlife goods if they purchase them in period 1—they return to the market in period 2 to resell those goods.

We will assume as before in the book that the consumers and producers of the wildlife goods are price-takers, i.e., they are all small relative to the size of the market; hence, each one has negligible influence on the market price of the good. In addition, for this section of the chapter, we will assume that the speculator sector is also competitive and that all speculators behave as price-takers as well. In a later section of the chapter, we will consider how our analysis and results differ if—instead of competitive speculators—there is a single, monopolistic speculator in the market who can significantly affect the market price.

Because there are two time periods in the model, there are two prices we need to keep track of: P_1, the price of the wildlife good in period 1; and P_2, the price of the wildlife good in period 2. The condition for the wildlife good market to be in equilibrium is that the prices P_1 and P_2 clear the market in both periods, i.e., supply must equal demand in periods 1 and 2.

Note that, when there are speculators in the market, total demand for the wildlife good in period 1 comes from the consumers as well as the speculators (while supply in period 1 comes only from the wildlife good producers). In period 2, however, demand originates completely from the consumers—since there is no benefit whatsoever for speculators to acquire the good when there are no future periods. Further, note that if speculators do acquire the wildlife good in period 1, then supply in period 2 comes from the wildlife good producers *and* the speculators.

If a speculator purchases the wildlife good in period 1, then—if the speculator is a price-taker with no market power—the speculator must choose to resell the good in period 2 since there is no benefit to holding on to the good in period 2 in a two-period world. Hence, the demand for the good in period 1 that comes only from the speculators must equal the supply of the good in period 2 that comes only from the speculators.[1] We will let $D_{SP}(P_1, P_2)$ denote the demand from speculators in period 1. Note that this speculator demand in period 1 should depend on both P_1 and P_2 since forward-looking speculators would take into account the period-2 price—which is the revenue generated from selling the wildlife good in period 2—when deciding whether or not to buy the good in period 1.

To state more formally the equilibrium condition for the market with speculators, let $D_1(P_1)$ and $D_2(P_2)$, respectively, denote the demand coming only from consumers in period 1 and period 2 given prices P_1 and P_2.[2] Analogously, let $S_1(P_1)$ and $S_2(P_2)$, respectively, denote the supply coming only from the wild-good producers in period 1 and period 2 given prices P_1 and P_2.[3] Market clearing in both periods requires that P_1 and P_2 satisfy the following:

$$D_1(P_1) + D_{SP}(P_1, P_2) = S_1(P_1),$$

$$D_2(P_2) = S_2(P_2) + D_{SP}(P_1, P_2).$$

Let us now consider the relationship between demand from the consumers in period 1 and the demand from them in period 2, as well as the relationship between the supply from wildlife-good producers in period 1 and the supply from them in period 2. If consumers' preferences and their willingness-to-pay for the wildlife good do not change from one period to the next, then there should be no difference between the demand curves in the two periods, i.e., $D_1(P) = D_2(P)$ for all P.

The relationship between the supply curves in the two periods is more subtle and complex, however. For one thing, the marginal cost of supplying the wildlife good—which determines what the supply curve is—should depend, among other factors, on the size of the wildlife population. In particular, the lower the population size, the harder it would be for poachers or hunters to locate the animals, and, hence, the more costly it would be for wildlife-good producers to supply one more unit of the wildlife good.

This implies that the supply curve of the wildlife-good producers in period 2 can depend on how much wildlife good is supplied in period 1. For example, if the equilibrium amount of the wildlife good supplied in period 1 is high, that would reduce the wildlife population size, increase the marginal cost of supplying the good in period 2 (compared to the marginal cost in period 1), and thus shift upward the supply curve of wildlife-good producers in period 2. On the other hand, if the amount of wildlife good supplied and harvested in period 1 is low, that could allow the population of the wildlife to grow. This, in turn, would lower the marginal cost of producing the wildlife good in period 2, thus shifting the period-2 supply curve downward.

For the purposes of our exposition below, we will assume that the population of wildlife under consideration is threatened or endangered in the following sense: If there were no speculators in the market, the equilibrium amount of wildlife good supplied in period 1 would cause the population to decline. This increases the marginal cost of supplying the wildlife good in period 2 (relative to the marginal cost in period 1) and shifts the period-2 supply curve upward.

Assume for now that the speculators do not discount future revenue (i.e., a one-unit payoff in period 2 is worth—in the speculators' eyes—the same as a one-unit payoff in period 1) and that the speculators do not incur any cost of storing the wild good in period 1. Let us also assume that there is a large number of potential speculators in the market in the sense that if there is a profit opportunity from stockpiling and speculating, that would attract new speculators to come into the market and purchase the wildlife good. We will assume there is no cost of entry or exit for the speculators.

Given these assumptions, the market clearing conditions above imply that, in equilibrium, we must have $P_1 = P_2$, assuming there is no depreciation of the wildlife good going from one period to the next. This is the zero-profit condition for an equilibrium. The intuition for this result can be given as follows:

- If $P_1 < P_2$, then the profit to a speculator of buying a unit of the good in period 1 and selling it in period 2 is positive. This opportunity would attract new speculators into the market. The influx of new speculators would increase the demand for the good in period 1, which would drive up P_1. At the same time, the increased supply of the good in period 2—from new speculators selling their good in period 2—would drive down P_2. Hence, any situation in which $P_1 < P_2$ cannot be a market equilibrium.
- If $P_1 > P_2$, then there would be no speculators in the market, as buying a wildlife good in period 1 and reselling it in period 2 would yield a negative profit. Recall our assumption stated earlier that, even without any speculators in the market, the population of wildlife would decrease in period 2, leading to a higher marginal cost of supplying the wildlife good in period 2 and hence a higher supply curve in period 2. This implies then that, since the demand curves in periods 1 and 2 are identical and the supply curve in period 2 is higher than the supply curve in period 1, the market-clearing price in period 2 must be *higher* than the market-clearing price in period 1. This is a direct contradiction of the case we are considering. Therefore, it is impossible for $P_1 > P_2$ in equilibrium.

The above discussions have shown that the price of the wildlife good in the two periods must be equal in a market equilibrium (assuming there is no storage cost or discounting of the future by the speculators). What exactly this common equilibrium price is depends on the specifics of the market, in particular, the market demand curve for the wildlife good, and the marginal cost of supplying the wildlife good in the two periods.

Graphical Analysis

We will utilize a graphical approach in this subsection to examine how the wildlife good market works when there are speculators in the market. As discussed previously, we assume that the consumer demand curve is the same in the two periods and given by $D(P)$.

To provide a benchmark for the analysis to follow, we first examine the market equilibrium when there are no speculators.

Market with No Speculators

Consider Figure 7.1. The market clears in period 1 at price P_1. We assume, as mentioned above, that the wildlife population of interest is threatened or

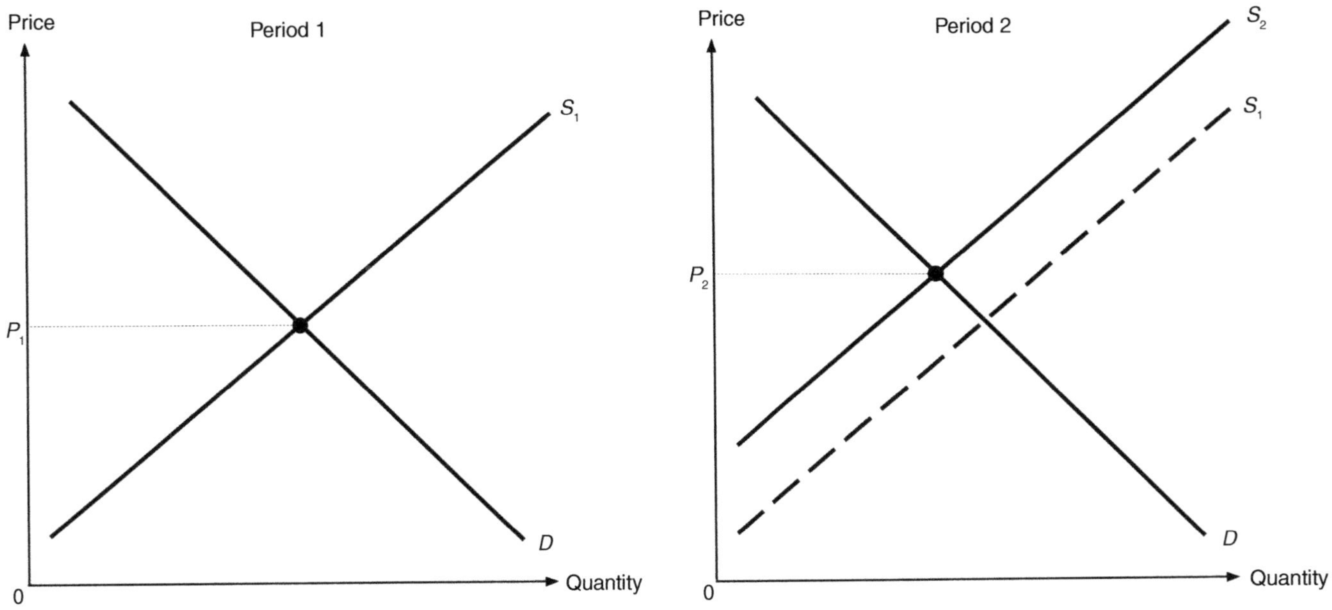

Figure 7.1 The equilibrium prices in the two periods in the absence of any speculators in the market. The supply curve in period 2 is higher than the supply curve in period 1 because of our assumption that—without any speculators—the equilibrium amount of wildlife good produced in period 1 would cause the wildlife population to decline in period 1 and in turn raise the marginal cost of production in period 2.

endangered so that—when Q_1 units of the wildlife good are supplied in period 1—the population declines in period 2, which raises the (marginal) cost of supplying the wildlife good in period 2. Therefore, the supply curve of wildlife good in period 2 is higher than the supply curve in period 1. Consequently, the market clears in period 2 at a higher price than the price in period 1, i.e., $P_2 > P_1$.

Market with Speculators

In Figure 7.1, because $P_2 > P_1$, if speculators were able to enter the market, they would take advantage of this arbitrage opportunity by buying the wildlife good in period 1 and reselling it in period 2. The additional demand in period 1 from the speculators would drive up demand and, hence, the price in period 1. A consequence of the higher amount of wildlife good sold in period 1 is that the supply curve of the wild-good sellers in period 2 would shift up even more (compared to the market without any speculators) because of a steeper decline of the wildlife population. While this factor alone would tend to increase the price in period 2, the additional supply of the wildlife good in period 2 from the speculators more than counteracts this factor and drives down the period 2 price. As long as the period 2 price exceeds the period 1 price, the aforementioned adjustments would take place until we reach the new equilibrium in which the two prices are equated. See Figure 7.2 for an illustration of how the new equilibrium prices are determined with speculators participating in the market.

Extensions

Storage Cost

Let us suppose now that speculators incur a cost of $\sigma > 0$ by storing a unit of wildlife good in period 1 so that it can be resold in period 2. With this storage cost, the profit to a speculator of buying and storing a unit of wildlife good in period 1 and reselling it in period 2 is thus $P_2 - P_1 - \sigma$, again, assuming no depreciation of the wildlife good. The zero-profit condition of equilibrium implies then that $P_2 = P_1 + \sigma$. Figure 7.3 shows graphically how the addition of this storage cost impacts the market equilibrium.

Discounting

Suppose speculators in period 1 discount period 2 revenue by the factor δ, where $0 \leq \delta \leq 1$. This means that a unit of payoff or revenue earned in period 2, from the perspective of period 1, is worth δ only. In this case, spending P_1 in period 1 on the wildlife good to resell it in period 2 yields a profit of, from the perspective of period 1, $\delta P_2 - P_1$. The zero-profit condition for an equilibrium then gives us $\delta P_2 = P_1$. This is illustrated in Figure 7.4.

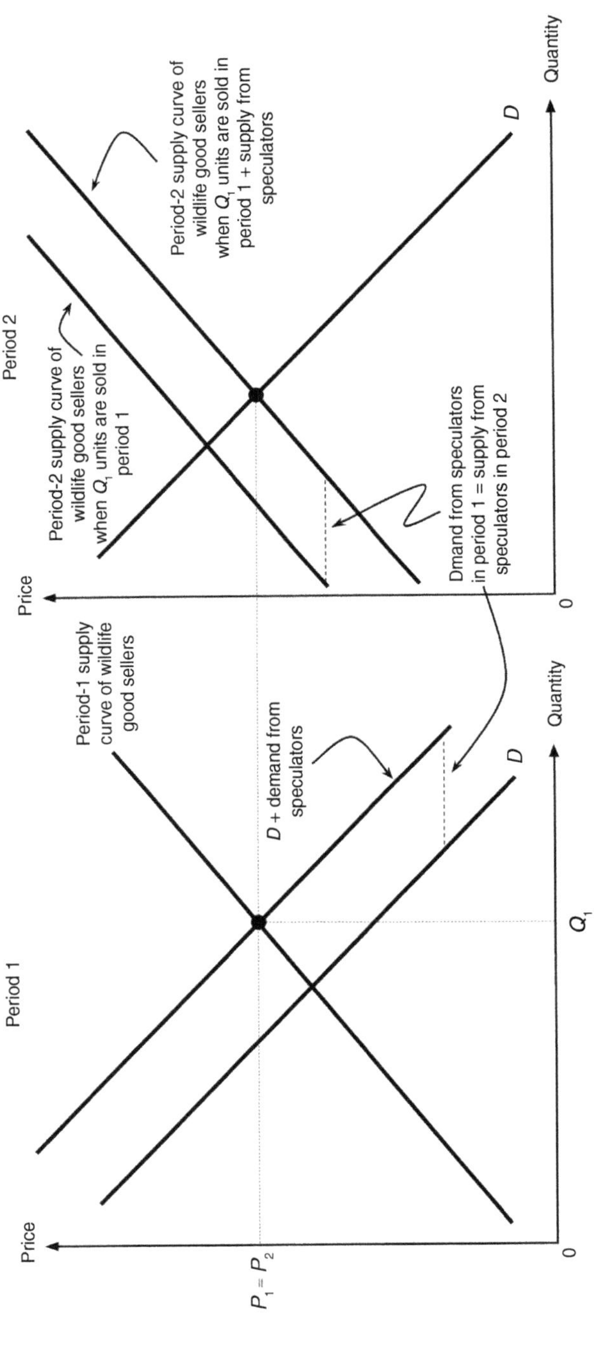

Figure 7.2 The equilibrium prices in the two periods with competitive speculators. Note that, compared to the situation without any speculators, more units of the wildlife good are produced in period 1 and more units of it are consumed in period 2.

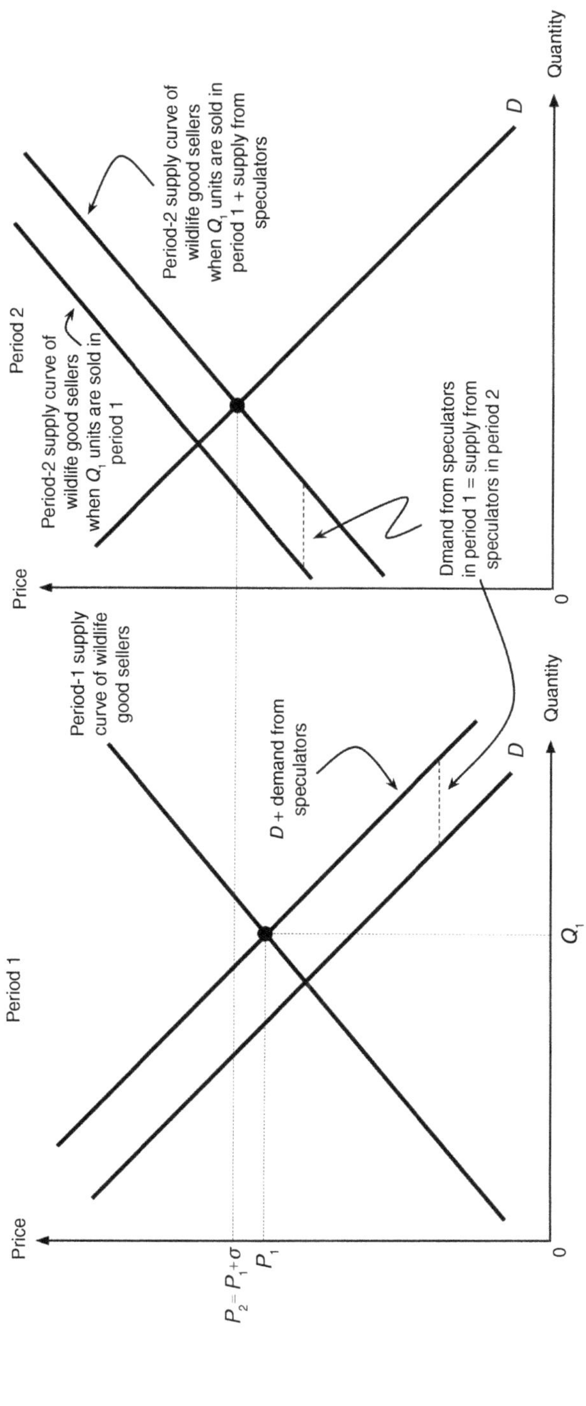

Figure 7.3 The equilibrium prices in the two periods with competitive speculators when the speculators have a storage cost of σ.

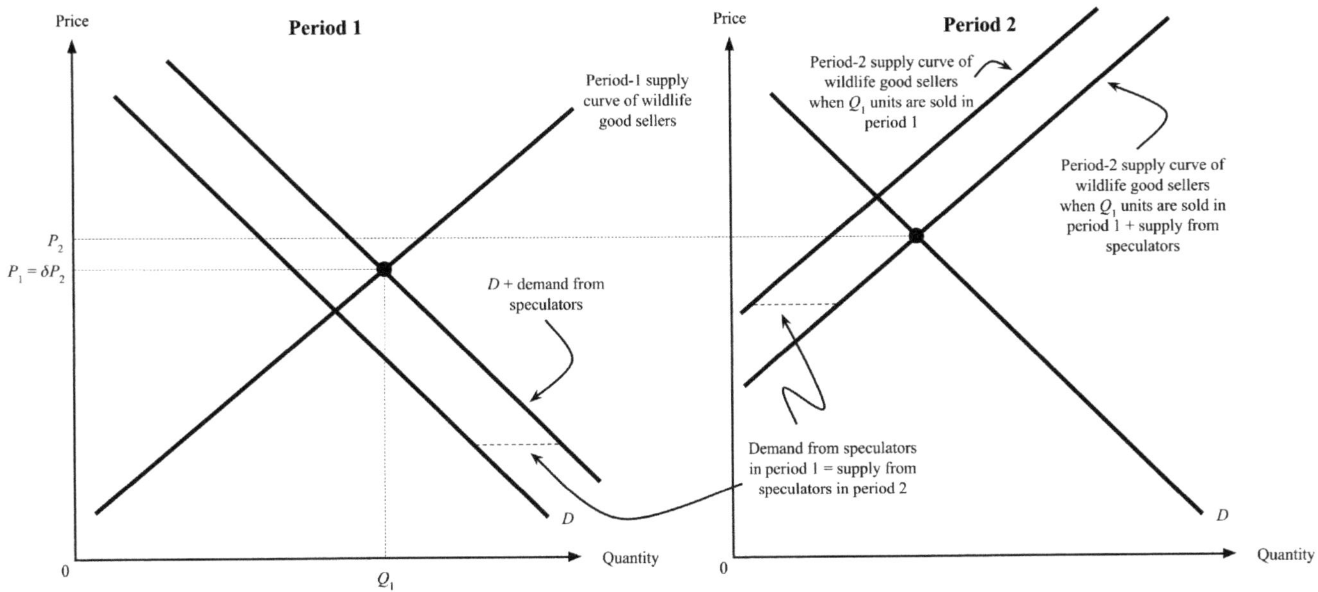

Figure 7.4 The equilibrium prices in the two periods with competitive speculators when the speculators discount future revenue by the factor δ.

Speculators with Market Power

In the previous section, we have examined the wildlife good market when each speculator is small relative to the entire market so that none of them—on their own—can exert significant influence on the market price. We have seen in this case that the equilibrium of the market is defined by the zero-profit condition, which—in the absence of storage cost or discounting by the speculators—implies that the price of the wildlife good is equated across time.

In this section, we deviate from the assumption that the speculators are price-takers and assume instead that there is a small number of speculators and that each is large relative to the market, i.e., each has market power and has significant influence on the market price of the wildlife good. In particular, we focus, for simplicity and concreteness, on the case in which there is a single speculator—i.e., the speculator is a monopoly—and it can stockpile multiple units of the wildlife good. As we shall see, the implications for wildlife trade of having a monopoly speculator in the market are markedly different from those when the speculator sector is perfectly competitive as assumed in the last section.

To understand intuitively why this is the case, note that the monopoly speculator—through its market power—has considerable control over the price of the wildlife good in periods 1 and 2. Assuming that the speculator's objective is to maximize its profit, there is no reason to expect the zero-profit condition to hold—we should expect the price of the wildlife good in period 2 to exceed the price in period 1.

In particular, by purchasing a large quantity of the good in period 1—thereby causing the wildlife population to decrease significantly and, as a result, raising the producers' marginal cost of supply in period 2 substantially—the monopoly speculator may be able to push the period-2 price to a higher level than is possible with a competitive speculator sector. Note that how high the monopoly speculator can push the period 2 price depends partly on how many units of the wildlife good it chooses to resell in period 2. Since the fewer units it resells, the higher the price is, *it may be to the monopoly speculator's benefit—i.e., from a profit maximization standpoint—not to resell all the units it purchases in period 1*. This is a key difference from the model with a competitive speculator sector. And one crucial implication of this is that more wildlife good could be supplied in period 1—i.e., the wildlife population could decline even more—when there is a monopoly speculator compared to when the speculator sector is competitive and composed of numerous price-taking speculators.

Numerical Example

To illustrate ideas, let us consider a concrete numerical example. We note that the example is not meant to represent or capture any particular wildlife good market—it is simply used here for expositional and pedagogical reasons. We also note that the example below is not exceptional or unique in that there is a myriad of similar examples that could be presented to illustrate the same points. All details and derivations for this example are given in the technical appendix for this chapter (available at: https://www.frederickchen.org/publications/the-economics-of-wildlife-trade-technical-appendices).

Consumers: On the consumer side, let us suppose that there are 80 consumers: 55 of them are low-value consumers who value the wildlife good at 150; and 25 of them are high-value consumers who value the wildlife good at 250. We assume for simplicity that each consumer can buy at most one unit of the good in each period.

Producers: On the producer side, suppose that there are 100 producers, each of whom can produce at most one unit of the wildlife good in each period. There are two types of producers: 50 of them are low-cost producers whose cost of production in period 1 is 100, and the other 50 are high-cost producers whose cost of production in period 1 is 200.

The cost of production in period 2 for each type of producer depends on the amount of the wildlife good that is supplied and sold in period 1, Q_1: for the low-cost producers, the period 2 cost of production is 180 if $Q_1 < 52$, and 280 if $Q_1 \geq 52$; for the high-cost producers, the period 2 cost of production is 260 if $Q_1 < 52$, and 360 if $Q_1 \geq 52$.

Speculator: There is a single speculator that we assume for convenience does not discount future payoffs and has no storage cost in the first period. Let σ_1 denote the quantity of the wildlife good the speculator purchases in period 1 and let σ_2 denote the amount of the good the speculator resells in period 2. Note that $\sigma_2 \leq \sigma_1$. Also note that, since supply is at most 100 in any period, $\sigma_1 \leq 100$.

Equilibrium: To find the market equilibrium for this example, let us consider the following possibilities.

- *Case 1*: $0 \leq \sigma_1 < 25$. In period 1, the market clears at price $P_1 = 150$, and the equilibrium quantity sold is $Q_1 = 50$.
- *Case 2*: $\sigma_1 = 25$. In period 1, the market clears at any price between 150 and 200, including 150 and 200. The equilibrium quantity sold is $Q_1 = 50$.
- *Case 3*: $25 < \sigma_1 < 75$. In period 1, the market clears at price $P_1 = 200$, and the equilibrium quantity sold is $Q_1 = 25 + \sigma_1$.
- *Case 4*: $\sigma_1 = 75$. In period 1, the market clears at any price between 200 and 250, including 200 and 250. The equilibrium quantity sold is $Q_1 = 100$.
- *Case 5*: $75 < \sigma_1 < 100$. In period 1, the market clears at price $P_1 = 250$, and the equilibrium quantity sold is $Q_1 = 100$.
- *Case 6*: $\sigma_1 = 100$. In period 1, the market clears at any price greater than or equal to 250, and the equilibrium quantity sold is $Q_1 = 100$.

We note that, since the price in period 2 can be at most 250—the willingness-to-pay of the high-value consumers—it cannot be profit-maximizing for the speculator to choose $\sigma_1 > 75$; hence, we can rule out Cases 5 and 6.

Now, consider Cases 1 and 2, i.e., $0 \leq \sigma_1 \leq 25$. In this scenario, the speculator in period 2 would try to target the high-value consumers and resell the good to them. However, it would not be able to get a price of 250 since the low-cost producers—with a period 2 cost of 180—would try to undercut any price above 180. Therefore, the period 2 market clearing price is $P_2 = 180$. The highest profit the speculator could potentially make in Cases 1 and 2 is thus $(180-150)(25) = 750$.

Consider Cases 3 and 4, where $25 < \sigma_1 \leq 75$. Note that if the speculator chooses to resell more than 25 units in period 2, then the period 2 price would be $P_2 = 150$ for the market to clear. If, however, the speculator resells 25 units or less in period 2, then the period 2 price would be $P_2 = 180$ or $P_2 = 250$, depending on how many units the speculator purchases in period 1. If $25 < \sigma_1 < 27$, then the highest period 2 price would be 180; if $\sigma_1 \geq 27$, then the highest period 2 price would be 250. Putting all these observations together yields the result that the highest profit is obtained in these cases when $\sigma_1 = 27$ and $\sigma_2 = 25$, and the maximum profit is $(250)(25)-(200)(27) = 850$. This exceeds the best that the speculator can do from cases 1 and 2, and this gives us a market equilibrium for this example: $P_1 = 200$, $P_2 = 250$, $Q_1 = 52$, $\sigma_1 = 27$, and $\sigma_2 = 25$.[4]

Note from this example that it is in the speculator's best interest to leave some units unsold in period 2. The reason this occurs is because it is worthwhile for the speculator to buy up "excess" stockpile in period 1 for the purpose of driving up the cost of production—and hence the price of the wildlife good—in period 2. This strategy is similar to what some scholars have called "banking on extinction.[5]

Equilibrium with Competitive Speculators

For the sake of comparison, let us consider what the equilibrium of this market is if—instead of a monopolistic speculator—there were many competitive speculators. We again assume that the speculators do not discount future payoffs and have no storage costs. (The derivations for this case are provided in the technical appendix for this chapter.)

For the zero-profit condition to hold, the prices in the two periods must be equal. It turns out that, for this example, there is a range of equilibrium prices, although there is a unique value for the equilibrium quantity bought and sold in each period. In equilibrium, 25 units of the wildlife good are purchased by speculators, which means a total of 50 units are sold in period 1. Any price between 150 and 180—including 150 and 180—is an equilibrium price for both periods. In addition, a total of 25 units of the wildlife good are sold in period 2, with all 25 units being provided by the speculators.

Compared to the example above with a monopoly speculator, note that the prices in both periods, as well as the quantity of wildlife good sold in period 1, are lower with competitive speculators. Since the quantity of the good sold in period 1 is a function of how much of the wildlife population is sacrificed to obtain the good, this example shows how critical the market structure of the speculative sector is from a wildlife conservation perspective.

The Great Elephant Ivory Stockpile Debate

We now consider how these insights into speculation might inform an approach to an ongoing conservation policy challenge: how to manage the world's stocks of elephant ivory for the benefit of *in situ* elephant conservation. This issue has been debated for decades, being placed into the spotlight in the late 1980s when African elephant poaching was out of control, and policymakers were trying to decide whether to ban the international ivory trade. To gain publicity and support for the ban, the newly appointed head of Kenya Wildlife Services (KWS), Richard Leakey, arranged for a 12-ton stockpile of confiscated ivory to be destroyed by being burned in a spectacular public event. Following this dramatic gesture, in 1989, the CITES Parties agreed to up-list African elephants in Appendix I.

Following the CITES decision, a group of economists argued that while it might provide short-term benefits as a moratorium (buying time to develop a more effective regulatory system), the trade ban would yield perverse results in the long term—and that establishing an appropriate ivory exchange system would be a better option (Barbier et al., 1990). In 1997, the CITES Parties agreed to allow three African countries to initiate what would later become the first of the two one-off sales (as discussed in Chapter 4). At the same time, they created a mechanism for donor countries and organizations to acquire ivory stockpiles for non-commercial purposes, raising the question of whether such stockpiles should be retained or destroyed ('t Sas-Rolfes, 1997).

The idea of stockpile destruction is contentious. While confiscating illegal ivory makes economic and conservation sense (as per the discussion on seizures in Chapter 4), destroying it afterward rather than selling it back into the market is a more complicated question. As economists have pointed out (e.g., Bergstrom, 1990), selling confiscated products is inadvisable in the case of harmful drugs for which the regulatory goal is to protect consumers by ensuring the drugs do not reach them; however, in the case of wildlife products, selling confiscated products is in theory sensible as it suppresses market prices and therefore the incentives for poaching (and the proceeds could ideally be reinvested into conservation, as a further benefit). The CITES approach, however, tends to follow a legalistic regulatory logic rather than an economic one: for products from Appendix I-listed species, sales of confiscated specimens are not permitted, and Parties are requested to either return them to source countries or otherwise store or destroy them (to avoid them potentially reentering illegal trade channels)[6]. This places an economic burden on the confiscating authority without providing an obvious immediate benefit for conservation.

Whereas CITES already started recommending the destruction of rhino horn stockpiles in the 1980s, most Parties ignored this, and some allowed them to accumulate in storage facilities ('t Sas-Rolfes, 2000). Similarly, many Parties started accumulating elephant ivory stockpiles, both from seizures and natural mortalities in range states, suggesting that some of their governments may be hoarding ivory for speculative purposes. The stockpile question attracted the attention of Harvard economist Michael Kremer (who later won a Nobel prize), who co-developed a model to consider links between storage and poaching (Kremer & Morcom, 2000) and concluded that governments should maintain stockpiles as an insurance policy against private speculative accumulation, which could drive harvesting toward extinction. In response, another group of economists cautioned that under some circumstances, profit-seeking governments might face incentives similar to a speculator with monopoly power, i.e., they could also help drive the species to extinction in the hope of earning a windfall profit from post-extinction sales (e.g., Mason et al., 2012).

Considering evidence that both governments and non-state actors have been stockpiling elephant ivory, what is the most sensible policy stance from a conservation perspective? The answer to this question depends somewhat on expectations of whether any future legal ivory trade will be permitted or not, and opinions differ on whether this is likely or desirable. In 2007, following approval of the second one-off sale, the CITES Parties agreed to a moratorium on any further sales until 2016. After 2007, governments followed diverging approaches, with some destroying stockpiles and others maintaining them, with southern African countries continuing to express interest in potential future sales. Contemplating three post-2016 options (maintaining stockpiles, selling them, or destroying them), stockpile destruction appeared to be the least precautionary and was advised against ('t Sas-Rolfes et al., 2014). However, this advice was ignored by Kenya, which staged another large-scale ivory burn in early 2016, this time of 105 tons estimated to be worth up to US$220 million (Biggs, 2016).

The Kenyan anti-trade stance and publicity again prevailed, gaining international support for further subsequent trade-restrictive approaches toward ivory (Braczkowski et al., 2018) and thwarting any attempts to reinstate a legal trading regime. In December 2016, the Chinese government announced a domestic ban on elephant ivory trade (with mammoth ivory trade remaining legal), prompting further discussion around the potential role of speculators, with recommendations that China commit to an indefinite ban and consider destroying its stockpiles (Harvey et al., 2017). However, China did not announce any subsequent stockpile destructions and elephant poaching persisted in parts of Africa after 2016 (Schlossberg et al., 2020). Statistical analyses of persistent poaching suggest that several economic and governance-related factors act as relevant drivers, including the market price of ivory (Hauenstein et al., 2019; Kuiper et al., 2023). The first empirical analysis of the impact of ivory destruction events on poaching rates in Africa revealed that they tended to raise both the ivory price and rates of poaching at the continental level (Gjerdseth, 2025).

If consumer demand for ivory can be expected to persist at sufficient levels despite continuing restrictions, speculators will continue to accumulate stockpiles. Our preceding theoretical analysis suggests that competition between multiple speculators will always result in lower prices and, therefore, lower poaching rates than if speculative stockpiling is monopolized (unless this happens under the banner of a non-profit collective committed to optimizing trade for the benefit of elephant conservation). Large-scale destructions of government stockpiles are thus likely harmful to elephant conservation, both because of their immediate price effects and because they reduce potential competition in the speculative marketplace. Unless and until global consumer demand for ivory disappears almost entirely (which seems highly unlikely in the foreseeable future), conservation-minded governments should probably avoid any further large-scale ivory stockpile destruction.

Notes

1 We will see in the next section that this statement does not necessarily hold when there is a single speculator with market power, i.e., it is possible when there is a monopolistic speculator for the speculator not to sell in period 2 all the wildlife goods that are acquired in period 1.
2 Unlike the demand from speculators in period 1, the demand from the consumers in period 1 is a function of only P_1 here. This is because, by assumption, the consumers cannot resell in period 2 what they purchase in period 1. Now, if consumers are able to save money, then it is possible for consumers to forego purchasing the wildlife good in period 1, save their money, and purchase the good in period 2 instead. In this case, consumers' purchasing decision in period 1 would depend on the period-2 price. For ease of exposition, we rule out savings (by which consumers effectively also become speculators) so that the demand in period 1 depends solely on P_1.
3 It is, in principle, possible for a wildlife good producer in period 1 to hold onto the good and wait to sell it in period 2. For simplicity, we rule out such a case and assume that the producers are "myopic."
4 Because there is a range of period-2 prices that clear the market when $\sigma_2 = 25$ —and yield higher profit than the other cases—there are in principle other market equilibria which differ from the given one in only the value of P_2. We present the equilibrium that yields the highest profit for the monopoly in the main text. Mathematically, as shown in Chen (2017), when there are multiple market-clearing prices, a market equilibrium for a monopoly is well-defined and exists if we pick the highest market-clearing price to work with.
5 See Mason et al. (2012).
6 The Parties to CITES recommend that:

> Parties dispose of confiscated and accumulated dead specimens of Appendix-I species, including parts and derivatives, only for bona fide scientific, educational, enforcement or identification purposes, and save in storage or destroy specimens whose disposal for these purposes is not practicable.
>
> (https://cites.org/sites/default/files/documents/COP/19/resolution/E-Res-17-08-R19.pdf)

References

Barbier, E. B., Burgess, J. C., Swanson, T. M., & Pearce, D. W. (1990). *Elephants, Economics and Ivory*. Earthscan.

Biggs, D. (2016). Track the impact of Kenya's ivory burn. *Nature, 534*(7606), 179–179. https://doi.org/10.1038/534179a

Braczkowski, A., Holden, M. H., O'Bryan, C., Choi, C.-Y., Gan, X., Beesley, N., Gao, Y., Allan, J., Tyrrell, P., Stiles, D., Brehony, P., Meney, R., Brink, H., Takashina, N., Lin, M.-C., Lin, H.-Y., Rust, N., Salmo, S. G., Watson, J. E. M., … Biggs, D. (2018). Reach and messages of the world's largest ivory burn. *Conservation Biology, 32*(4), 765–773. https://doi.org/10.1111/cobi.13097

Chen, F. (2017). The economics of synthetic rhino horns. *Ecological Economics, 141*, 180–189. https://doi.org/10.1016/j.ecolecon.2017.06.003

Do, Q.-T., Levchenko, A. A., Ma, L., Blanc, J., Dublin, H., & Milliken, T. (2020). The price elasticity of African elephant poaching. *The World Bank Economic Review, 35*(3), 545–562. https://doi.org/10.1093/wber/lhaa008

Gjerdseth, E. A. (2025). Crush and burn: How the destruction of ivory fails to save elephants. *World Development, 185*, 106766. https://doi.org/10.1016/j.worlddev.2024.106766

Harvey, R., Alden, C., & Wu, Y.-S. (2017). Speculating a fire sale: Options for Chinese authorities in implementing a domestic ivory trade ban. *Ecological Economics, 141*, 22–31. https://doi.org/10.1016/j.ecolecon.2017.05.017

Hauenstein, S., Kshatriya, M., Blanc, J., Dormann, C. F., & Beale, C. M. (2019). African elephant poaching rates correlate with local poverty, national corruption and global ivory price. *Nature Communications, 10*(1), 2242. https://doi.org/10.1038/s41467-019-09993-2

Kremer, M., & Morcom, C. (2000). Elephants. *The American Economic Review, 90*(1), 212–234.

Kuiper, T., Altwegg, R., Beale, C., Carroll, T., Dublin, H. T., Hauenstein, S., Kshatriya, M., Schwarz, C., Thouless, C. R., Royle, A., & Milner-Gulland, E. J. (2023). Drivers and facilitators of the illegal killing of elephants across 64 African sites. *Proceedings of the Royal Society B: Biological Sciences, 290*(1990), 20222270. https://doi.org/10.1098/rspb.2022.2270

Mason, C. F., Bulte, E. H., & Horan, R. D. (2012). Banking on extinction: Endangered species and speculation. *Oxford Review of Economic Policy, 28*(1), 180–192. https://doi.org/10.1093/oxrep/grs006

Schlossberg, S., Chase, M. J., Gobush, K. S., Wasser, S. K., & Lindsay, K. (2020). State-space models reveal a continuing elephant poaching problem in most of Africa. *Scientific Reports, 10*(1), Article 1. https://doi.org/10.1038/s41598-020-66906-w

't Sas-Rolfes, M. (1997). Elephants, rhinos and the economics of illegal trade. *Pachyderm, 24*, 23–29. https://doi.org/10.69649/pachyderm.v24i1.900

't Sas-Rolfes, M. (2000). Assessing CITES: Four case studies. In J. Hutton & B. Dickson (Eds.), *Endangered Species, Threatened Convention* (pp. 69–87). Earthscan Publications.

't Sas-Rolfes, M., Moyle, B., & Stiles, D. (2014). The complex policy issue of elephant ivory stockpile management. *Pachyderm, 55*, 62–77. https://doi.org/10.69649/pachyderm.v55i.355

8 Influencing Actor Incentives in Wildlife Trade

Having laid out our basic economic model of the market for wildlife goods—and its various extensions—and having further considered the implications of specific policy interventions, we now take a different angle by considering how this knowledge could be applied to approaching wildlife trade strategically to benefit conservation. We consider this question from the broad perspective of any person or organization concerned with *in situ* species conservation, recognizing that certain policies, regulations, and actions can often only be undertaken by specific parties. We also consider potential implications for the welfare of the wildlife concerned. We present our approach in two parts over the next two chapters. In this chapter, we build on our description of market actors from Chapter 3 to consider how their incentives may be determined and potentially influenced. In Chapter 9, we combine these insights with those from Chapters 4–7 to consider the most appropriate ways to shape trade policy and regulation for the benefit of conservation. We start this exposition on the supply side by considering the perspective of those directly responsible for the custodianship and management of wildlife.

Supply-Side Incentives

The supply side of wildlife trade chains is the most critical—after all, this is where actual conservation takes place. We consider the primary goal to pursue conservation objectives as laid out by the IUCN and other prominent mainstream environmental organizations, i.e., to conserve genetically viable populations of wild species within their natural habitats and functional ecosystems, where they remain subject to natural evolutionary processes. As discussed in Chapter 2, there is some debate over which specific aspects require the most focus to achieve effective *in situ* species conservation, but there is at least widespread agreement that humanity should strive to avoid causing species extinctions in the wild. Hence, our focus here is to ensure that humans do not exploit wild species at unsustainable rates of extraction—to the extent that wild animals, plants, and fungi are removed from the wild, this should take place at a rate that does not exceed the potential of populations

DOI: 10.4324/9781003286318-8

to continually withstand such harvesting. Given this broader social objective, we need to look more closely at the constellation of actors on the supply side to understand their mix of incentives. Within the scope of "conservation," we can identify at least two distinct actions: protection and harvesting. The challenge for conservationists[1] is to establish and maintain the optimal mix of both.

Optimal Harvesting

The question of what constitutes an optimal rate of harvesting is the subject of an interdisciplinary literature on "sustainable use" (e.g., Milner-Gulland & Rowcliffe, 2008), which draws on species ecology and combines insights from population dynamics with economics to derive bioeconomic models of optimal harvesting.[2] Such models require input from scientists who understand the biophysical elements of species survival and reproduction, employing knowledge from various strands of biology. The biological characteristics of a species will somewhat determine the extent to which it can be harvested sustainably. With reference to fisheries, cautious biological scientists have highlighted that the population dynamics of many wild species may be poorly understood, expressing general pessimism that any routine harvesting can be sustainable in the long run (e.g., Ludwig et al., 1993); others have disagreed and expressed more optimism (e.g., Rosenberg et al., 1993). However, as discussed in Chapter 2, rates of harvest by humans are not determined by biological factors alone; they will also be strongly influenced by institutional arrangements and economic incentives.

A more detailed discussion of the biological factors influencing the potential supply of particular species is beyond the scope of this book; however, two points are worth mentioning. The first is that reproductive and population growth rates vary significantly, not only between species but also within species, depending on biophysical and other contextual factors that affect living conditions. Many species population growth rates are density-dependent, meaning that they are affected by population size relative to a defined area. At very low or very high densities, population growth rates may slow down—at low densities because individuals struggle to find a mate and at high densities because of increased resource competition that reduces fertility and raises mortality rates of new progeny. Hence, most species will have an optimal "in-between" density range within which the reproductive success rate (and therefore population growth rate) will be the highest.

The second point worth noting is that bioeconomic models focused only on optimal extraction typically overlook the fact that *in situ* species populations often provide value to humans beyond their mere harvest value. Many species provide (sustainable) use values while alive, including economic benefits of directly non-extractive enjoyment (e.g., through tourism viewing or media consumption) and indirect existence value, which may be realized in ways that can potentially be channeled into protective action.[3] The bioeconomics literature predicts that species with an economic harvest value may be harvested to extinction if their intrinsic population growth rates are lower than an applied economic discount rate (Clark, 1973).[4] However, as indicated in Chapter 2, such overharvesting is most likely to occur when genuine open access conditions exist, and this can be mitigated by establishing stronger property rights. With stronger property rights, the custodians of economically valuable species typically have greater incentives to protect them against overharvesting, even if their population growth rates are slow.

Incentives to Conserve

In the supply chain, the actors responsible for conserving a species face a different set of incentives from other actors. Conservation actors must normally prioritize protection over harvesting, ensuring that the latter only occurs at sustainable rates of extraction. The principal conservation actors are the custodians of wildlife and their habitat and typically comprise government agencies, which may or may not bear the direct responsibility for management through ownership and control.[5] However, not all wildlife is found on public land—in many parts of the world, it also occurs on communal and private lands. In such circumstances, the jurisdiction over—and incentives for—wildlife management may vary. A detailed discussion of all the possible institutional arrangements and their implications is again beyond the scope of this book, but two related strands of economic theory are worth considering to gain a better understanding of institutional strength and incentives to conserve effectively. The first is the property right paradigm (Alchian & Demsetz, 1973) and its subsequent refinements that were discussed in Chapter 2. The second is principal–agent theory (Laffont & Martimort, 2009), including the concept of the residual claimant (Fama & Jensen, 1983; Williamson, 1983).

Principal–agent theory examines potential conflicts of interest and variations in priorities that occur when one party (the "agent") represents and acts on behalf of another (the "principal"). The core intuition of this body of theory is that the further removed the principal is from the action being taken by the agent, the greater the disparity in the information they possess (similar to the asymmetric information issue discussed in Chapter 6) and the probability that their interests and priorities will diverge. A person or entity directly owning and controlling wildlife is most likely to manage it according to their own interests. When people rely on another agency or even a chain of agencies to represent their interests in conservation, there is a greater risk that their objectives will diverge.

Not only is the effective distance between the principal and agent relevant, but in any collective entity, the number of ultimate principals (residual claimants) also matters. The intuition here is that with a high number of residual claimants, the benefits are so dispersed that none has a strong individual incentive to ensure that the agent effectively represents their interests.

This type of free-rider problem is essentially similar to the so-called "Tragedy of the Commons."

These insights from property rights and principal–agent theory predict that, for example, private rhino owners will, on average, perform far better in protecting animals (from poaching) that they own and control on their land than government agencies appointed to protect rhinos on public land.[6] For rhino conservation, the theory is indeed supported by substantial empirical evidence in Africa—decentralization of rhino management has substantially outperformed centralized state control since the early 1970s; African governments that did not devolve any authority lost all their rhinos to poaching ('t Sas-Rolfes & Emslie, 2025). Applying these theoretical insights more broadly to wildlife custodianship and conservation management suggests that incentives to conserve effectively may be increased significantly through greater devolution of responsibility.[7] However, many conservation scientists are skeptical of the role of non-state actors, especially private landowners, in conserving wildlife (Gooden & 't Sas-Rolfes, 2020).

As a final critical point, custodians of wildlife with harvestable value require resources to physically protect it from overexploitation. In the case of high-value species, the incentives for commercial poaching are high, and conservation agencies depend on adequate and sustainable sources of funding to acquire and maintain the necessary human resources, equipment, and technology to mitigate such activity. Agencies with insufficient resources are unable to prevent poaching or even adequately protect their own staff, which lowers morale and increases risks of internal corruption (again further aggravated when principal–agent factors are more salient). A lack of financial resources on the ground (at the source of wildlife products) remains a major challenge for conservation in many parts of the world, especially in the Global South. In some instances, a carefully managed regime that enables sustainable harvesting of wildlife products may assist with providing additional resources.[8]

Incentives to Harvest

Motivations to harvest wildlife products vary, but consumption and trade are dominant factors. Large-scale harvesting of the type that threatens species is typically economically motivated. We can thus model the incentives to harvest using a basic harvester's profit function. First, we assume that any potential harvester faces an option to engage in some sort of harvesting activity and another (best) alternative to allocate their time and resources. If the net benefit to harvest exceeds the net benefit of the alternative, we can expect them to choose the option to harvest. We can represent this as follows: $\Pi_h > \Pi_a$, where Π_h is the profitability of harvesting and Π_a is the profitability of the alternative. We can further represent the profitability of harvesting as being the benefits of harvesting minus the costs. In the simplest case, we represent this as $\Pi_h = B - C$, where B and C are the benefits and the costs of harvesting, respectively.

Assuming that only one type of wildlife product is being harvested, the benefits of harvesting *B* can be further broken down into two components: *P*, the selling price[9] of the product, and *Q*, the number of units harvested. If harvesting is legal, the profitability can simply be expressed as follows: $\Pi_h = PQ - C(Q)$, where *C(Q)* represents the total cost of harvesting *Q* units, including components such as equipment needed to harvest, travel costs, etc. However, if harvesting is illegal, a further cost factor must be added, namely, the expected value of any penalty if the harvester is apprehended and punished. This value may be expressed by the compound variable of ΘF in which *F* represents the economic value of the penalty and Θ represents the perceived probability of incurring it. For the illegal harvester, therefore, the profitability of harvesting can be expressed as follows:

$$\Pi_h = PQ - C(Q) - \Theta F$$

What is the relationship between the rate of harvesting and its profitability? All things being equal, we would associate increased profitability with an increased incentive to harvest.[10] In other words, there would be more harvesting effort, both by any individual and by a greater number of individuals, who would increasingly be attracted to this activity relative to their alternative sources of income. If the harvesting effort increases, we can also expect the harvest rate to increase (at least initially).[11] Therefore, if our conservation objective is to slow down the rate of harvesting, we need to consider ways to reduce the relative profitability of harvesting. This can be done in several ways:

- By increasing the value of the alternative, Π_a,
- By reducing the value of the product price, *P*,
- By raising the per unit costs of harvesting, *c*,
- By increasing the perceived probability of incurring a penalty, Θ, and
- By increasing the economic value of the penalty, *F*.

Observant readers will notice that each of these five options can be linked to the supply and demand visualizations provided by our basic economic model in Chapter 4. In each case, the direct or indirect effect of these interventions would be to reduce the quantity *Q* of wildlife product being traded, which amounts to an effective reduction in harvest. We now consider each of these five potential interventions in more detail.

Alternative Livelihoods

As discussed in Chapter 4, providing more appealing livelihood opportunities to illegal harvesters could reduce their incentives to attempt poaching. This idea of providing alternative livelihoods to local people as a conservation intervention gained traction during the latter twentieth century, prompting a range of new projects focused on community-based conservation, community-based natural resource management, and integrated coastal zone management (Wright et al., 2016). While the theoretical basis for this approach appears sound, there are numerous challenges with its implementation in practice, and a comprehensive global assessment of alternative livelihoods projects through a systematic literature review in 2015 generated disappointing results. Of 106 identified projects, only 21 had attempted to measure conservation effectiveness, and only nine had reported that the intervention was somehow effective, mostly by indirect assessment of attitudes and behavior (Roe et al., 2015). Only one project reported a direct positive impact on the population of an affected species.

The alternative livelihood approach assumes that simply providing more income than hunting or poaching would reduce those activities. However, the evidence to date suggests that this assumption is seldom correct in practice and that the provided alternative sources of income can simply become supplements rather than replacements for income from resource exploitation (Wright et al., 2016; Torell et al., 2010). A study on meat hunting in Ghana demonstrated that hunters already had other income streams from agricultural activities but chose to hunt seasonally to optimize the time required to attend to commitments related to those other activities (McNamara et al., 2016). Much of the research also reveals that hunting and other forms of harvesting are partly driven by strong traditional and cultural factors, not only commercial ones.

Following these assessments, conservation scientists and practitioners recommended improved approaches to alternative livelihood interventions. These improvements include better pre-implementation research, developing a "theory of change" (Biggs et al., 2017), specifying the intended species conservation outcome and establishing an appropriate monitoring system, conducting a risk assessment, and generally applying social-ecological systems thinking to contextualize the intervention within a broader systems map that integrates socio-cultural factors. Alternative livelihoods strategies may comprise necessary (but not sufficient) measures to combat poaching in certain areas, but they need to be carefully designed and implemented and, unless very clearly targeted, are only likely to yield general improvements over longer time horizons.

Suppressing Product Prices

In contrast with other approaches, measures to suppress product prices aim directly at reducing the price a harvester may expect to receive from a product buyer. Note that this approach is not always unambiguously desirable—as discussed earlier, there are instances in which species conservation may benefit from the income received from sustainable wild harvesting and product sales. These include situations in which such income is reinvested into direct species conservation measures and essential habitat management, and those in which it may be distributed to local communities to secure their cooperation in conservation initiatives (as with, e.g., legal black rhino hunting in Namibia). In such situations, it may be optimal to harvest lower (sustainable) quantities and achieve the highest possible price per unit. However, there are many instances in which the necessary institutional arrangements for appropriate benefit capture and resource allocation are missing (e.g., for open-access marine species), and for these, it might be desirable to reduce the prices of their wild-harvested products.

In theory, there are several ways to reduce wildlife product prices. Our economic model shows us that this can be achieved through measures that reduce demand for the affected species' wild-harvested product or otherwise increase supply, either of the product itself (from sustainable harvesting or accumulated stockpiles) or of a suitable competitive substitute. As previously discussed, trade restrictions or bans may or may not reduce the underlying consumer demand (i.e., there may be no shift in the demand curve), depending on its nature and primary motivations. If legal restrictions fail to reduce this demand, they may cause price increases rather than reductions and thereby stimulate increased poaching efforts. Therefore, a better approach to reducing demand is to appeal directly to consumers through measures such as public information campaigns, social marketing, or direct contact (Veríssimo 2013; 2019). We discuss demand reduction measures further below in the section on demand-side incentives.

Increasing supply from an alternative sustainable source—and ideally influencing consumers to switch to products from that source—provides another way to suppress or reduce prices. As discussed in Chapters 5 and 6, this can be achieved in several ways but needs to be approached with some care. An ideal scenario would be the introduction of a distinct but more affordable product that is at least partly accepted as a substitute. Such an approach would be most likely to reduce the demand for the competing wild product, leading to lower prices and incentives for illegal suppliers (poachers). However, it would not remove the need for anti-poaching measures, and the conservation benefits would be somewhat undermined if the substitute product is aggressively marketed to achieve a higher price for it after it is introduced (thereby risking spillover effects onto the price of the wild product).

If there is an option of introducing a substitute product that is essentially indistinguishable from the wild product, other factors must be considered. If trade in the wild-harvested product is already regulated or banned, introducing such a product will raise concerns about creating potential laundering opportunities. The validity of such concerns may be questionable from a conservation economist's perspective (if, e.g., there is a real possibility of flooding the market with low-cost inferior indistinguishable substitutes, thereby provoking an adverse selection response), but risk-averse regulators may be inflexible (as is currently the trend within the CITES community) and may not permit the sales of such products. As indicated in Chapter 6 and elsewhere (Chen & 't Sas-Rolfes, 2021), there are circumstances under which the use of a permit system for legal sales would result in a price reduction effect of the wild-harvested good that exceeds the downside of any potential laundering effect. However, any attempts to implement this approach would need to be carefully evaluated to assess the magnitude of all marginal effects.

An interesting variant of price suppression occurs when a wildlife manager reduces the realizable value of a wild specimen by preemptively harvesting the most valuable portion from it while it is alive (and before a poacher can reach it). One example of this approach is rhino dehorning, which is now widely used in Africa to deter rhino poaching. Because poachers are paid for rhino horn by weight and most of the horn volume can be removed without harming the animals, a recently dehorned rhino has considerably less value to a poacher than one with a full horn. Dehorning is not a panacea, as it is costly, and horns regrow (so that its deterrent effects wear off)[12]; nevertheless, a comprehensive study found it to be the most effective method of protecting rhinos in the vicinity of South Africa's Kruger National Park (Kuiper et al., 2025), confirming the significance of price as a variable that influences the incentive to poach.

Raising Costs of Harvesting

As with the rationale for suppressing wildlife product prices, raising the costs of harvesting may not be unambiguously desirable, depending on whether the harvesting in question is deliberately managed to benefit conservation or not. The most obvious way to raise the cost of harvesting is through various forms of restrictive regulation. Such regulation may be aimed at restricting the times and places at which harvesting can take place or the methods used for harvesting.[13] Note that all such regulations may be ignored, but they do have the effect of imposing costs on the harvester, including the added expected value of a penalty should they ignore the regulations. Other ways to raise costs are through physical deterrents such as creating distance or barriers such as fences and checkpoints along travel routes, all of which raise the costs of movement to and from areas with wildlife. Further indirect measures may include restrictions on the general availability of equipment used for harvesting (e.g., certain types of firearms and other weapons, as well as trapping devices). For many species, however, the scope for raising harvesting costs may be limited, especially outside of designated access-controlled protected areas.

Increasing the Perceived Probability of Punishment

As mentioned in Chapter 4, more effective anti-poaching measures can raise both the actual and perceived cost of illegal harvesting, thereby reducing the harvesting rate. The effectiveness of such measures depends on the expected value of a potential penalty being incurred. This, in turn, depends on the effectiveness of a range of anti-poaching measures. As explained in the next section, penalties for poaching vary considerably across species and jurisdictions. Similarly, the probabilities of incurring such penalties also vary greatly with circumstance. Both the economic literature on crime (e.g., Becker, 1968) and key scholars in criminology (e.g., Nagin, 2013) affirm that the probability of incurring a penalty is a far more significant factor in deterrence than the value of the penalty itself, which may be heavily discounted if the probability of incurring it is low. It is also worth noting that the deterrent incentive is not the actual probability (as empirically determined) but rather the perceived probability, which may be lower when prospective poachers have limited access to information or when they are young or desperate and inclined to discount the future at a higher-than-average rate.

From a conservation perspective, the probability of early detection and apprehension before an actual poaching attempt is a critical deterrent, especially when poaching is lethal. The conservation value of after-the-fact detection and apprehension is significantly lower, given that the wild organism has already been lost and given that arrest, successful prosecution, and effective punishment are all costly and time-consuming activities to conduct. From the perspective of the potential poacher, the probability of early detection and interception is also critical. In some instances, in certain jurisdictions, anti-poaching units may resort to lethal measures ("shoot-to-kill" policies—see Messer, 2010; Mogomotsi & Madigele, 2017), but poachers frequently continue to risk their lives, suggesting that they perceive the probability of being intercepted and killed as sufficiently low to warrant the risk. Where such lethal measures are preferably avoided by enforcement agents, the probability of incurring a penalty is significantly reduced, as it amounts to a compound probability of a sequence of events— detection, interception, arrest, prosecution, and full punishment—all of which must take place effectively without being thwarted.

Increased Value of Penalty

Increasing the value of the penalty for poaching is often advocated as a solution. However, not only is this measure potentially undermined if the perceived probability of incurring the penalty is low, but it is also challenging to determine penalties that accurately reflect the social cost of lost value that takes place through legal infractions. If trade is legal, poaching penalties can be linked to the value of the poached wildlife specimen. However, if trade is completely illegal, as rhino horn has been for several decades, prices are clandestine and somewhat inconsistent, and it is unclear whether penalties should be linked to the fees paid to poachers or recorded wholesale and retail prices in consumer markets, which are multiple times higher. Another approach is to link penalties to foregone income opportunities from activities such as tourism viewing or other assessed economic values such as the provision of ecosystem services.

A survey across various rhino range states reveals a wide range of prescribed penalties for rhino poaching, with fines ranging from a minimum of US$62 to a maximum of US$ 3.3 million, and prison sentences ranging from 1 month to life imprisonment (see 't Sas-Rolfes et al., 2025: 15, Table 15.1). While the lowest penalties almost certainly provide an insufficient deterrent, the effect of extremely high penalties may also be undermined in jurisdictions in which the public may consider them to be excessive. In such instances, a severe penalty may reduce the probability of successful arrest, prosecution, and full punishment, as local enforcement officials may be reluctant to abuse others in their community, and judges may require higher standards of evidence before imposing such extreme measures, thereby reducing the expected value of the penalty and undermining its effect. Most poachers tend to be poor, marginalized people, unable to afford high fines, and thus more likely to require imprisonment, which places a substantial socio-economic burden on their families and society in general. Increased penalties, therefore, have limitations, especially when the probabilities of incurring them are perceived to be low. A high probability of preemptive interception remains the most effective deterrent for poachers.

Demand-Side Incentives

As discussed in Chapter 3, consumers play a critical role in driving wildlife markets. Ideally, their behavior should be influenced such that their purchasing decisions support conservation; however, this may not always be possible, in which case they will simply respond to prices and other economic incentives. Depending on the wildlife product, one or more of four different strategies could be adopted to influence consumer behavior:

1 Encourage purchases of sustainably sourced (legal) products while discouraging purchases from other (potentially illegal and unsustainable) sources.
2 Encourage purchases of more affordable substitute products.
3 Discourage purchases of the product using taxes and other measures to reduce consumption levels.
4 Discourage all purchases through domestic trade bans and demand reduction measures.

Option 1 could be pursued by legal sellers (who would have an incentive to do so) but may be vulnerable to the laundering problem and would ideally require some form of product traceability and certification mechanism to be in place. It follows the logic of lowering the costs of the preferred option relative to the undesirable alternative (illegal product). Option 2, which follows a similar logic (as outlined in Chapter 5), is aimed at reducing the price that consumers would be willing to pay for the target wild product and could be achieved by somehow subsidizing the price of the substitute product. Option

3 is an approach used for products such as cigarettes and alcohol, for which a combination of public messaging and special taxes is used to discourage high consumption levels. There are some limitations to this approach—for example, if taxes are excessive, this may stimulate more concealed illegal trade.[14]

Option 4 depends in part on law enforcement, but it is challenging to enforce trade bans at the consumer level. While it may be possible to monitor physical retail outlets, much of the contemporary wildlife trade has moved online, where it is much harder to monitor and police (Soriano-Redondo et al., 2023). This approach must, therefore, also rely on direct appeals to the public to stop buying the product by creating a stigma against its use. For such direct appeals to be effective, they must be culturally sensitive and appropriate. A review of campaigns targeting rhino horn consumers in Vietnam revealed many shortcomings (Dang & Nielsen, 2021). A broader systematic review of conservation interventions to promote voluntary behavior change also demonstrated a strong geographic implementation and study bias toward North America and Europe (Thomas-Walters et al., 2023a). Successful implementation of consumer demand reduction initiatives in most key wildlife product markets (e.g., in Asia) remains largely elusive. To the extent that some consumer behavior change has been observed, it typically takes a long time and involves multiple factors—for example, the gradual decline of elephant ivory demand in Japan (see Thomas-Walters et al., 2023b).

Intermediary Incentives

Since intermediaries are typically entrepreneurs and other actors seeking to earn a profit, their incentives can be understood in much the same way as commercial harvesters, i.e., by applying a similar profit function analysis to the one specified above. In this instance, it may be challenging to specifically provide them with better economic alternatives, given that almost anyone could act as an intermediary, and most intermediaries will already have alternatives. Accordingly, all efforts should be focused on the profitability of intermediary activities, including trading, transporting, and stockpiling. It is, however, also challenging to influence product prices directly within these activities, given that prices are ultimately driven by consumer demand. Most efforts should thus focus on raising the costs (and risks) of all intermediary activities related to illegally and unsustainably harvested products, notably product laundering and smuggling. As with harvesting, establishing and enforcing appropriate (i.e., incentivizing) regulations provide the clearest way in which to achieve this.

The best way to regulate any specific wildlife product supply chain will vary with the product in question, and predominantly legal markets require different approaches to those for essentially banned products—we discuss this further in Chapter 9. In all cases, however, intermediary activities within supply chains can be challenging to monitor and control, with the best opportunities to do so being when products cross enforceable boundaries such as international borders. As with poaching, increasing the expected penalty for illegal intermediary activity is theoretically desirable but challenging, given the relatively low probabilities of detecting smuggling activity and substantially lower probabilities of dismantling professional smuggling syndicates. Nevertheless, stronger law enforcement measures, including harsher penalties, product seizures, and asset forfeiture for sufficiently serious cases, may all act as deterrents for all but the most organized of criminals.

Finally, incentives for speculative stockpiling should be carefully considered, especially where this may contribute toward accelerated harvesting rates. Stockpiling of wildlife products may take place at an individual level or at scale, with the latter being significantly riskier for intermediaries if the products are illegal. However, as discussed in Chapter 7, opportunities for monopoly profits may incentivize organized crime syndicates to do this. Speculative stockpiling can be discouraged by lowering expectations of future profits, which may be achieved through measures to reduce consumer demand, raise the cost of illegal trading, or increase competition in the marketplace (e.g., by introducing new competing sources of supply into the market).

Integrated Approaches

Considering incentives along entire trade chains reveals the potential for conflicting measures. For example, sustainable legal trading regimes that support conservation may be undermined by demand reduction initiatives. Excessive regulations that are inadequately implemented may stimulate the growth of illegal supply chain networks that evolve to become more efficient at (unsustainable) extraction than legal suppliers and difficult to dismantle. Destruction of legal stockpiles in one jurisdiction may raise the incentives for illegal speculative stockpiling and accelerated harvesting in others. When considering how to approach wildlife trade for the benefit of conservation, it is essential to apply systems thinking and adopt an integrated "all-of-trade-chain" approach. Unfortunately, the current international wildlife trade regime framed by CITES fails to embrace such inclusivity with its narrow regulatory approach (Challender & McMillan, 2014; Cooney et al., 2021). Not only does this approach appear to be working poorly for many species, but the structure and mechanism of CITES appear to lock in trade policy conflict, especially given tensions between influential interest groups pursuing somewhat incompatible agendas within the CITES system ('t Sas-Rolfes & Gooden, 2024; 't Sas-Rolfes et al., 2025).

In this chapter we have discussed some of the ways in which the incentives of actors involved in wildlife trade may be influenced for the benefit of conservation. Linking this socio-economic understanding with an ecological understanding of what constitutes sustainable use of a species provides a social-ecological systems context in which trade policy can be considered holistically. Applying such thinking to wildlife trade regulation could significantly improve its performance (Challender et al., 2025). In the next chapter, we take a closer look at strategic ways in which future wildlife trade regulation could be approached, having adopted the lessons from our economic models and their implications.

Notes

1. We use the term "conservationists" here in the way in which we believe it was intended when first introduced in the early twentieth century, as distinct from "protectionists" who may oppose any extractive harvesting by humans of one or more wild species as a matter of principle (due to concerns about animal welfare or ideological opposition to the commodification of wildlife products).
2. Note that the term "sustainable use" can be somewhat controversial and interpreted in different ways (see Hutton & Leader-Williams, 2003). Nevertheless, it appears as one of three central objectives of the Convention on Biological Diversity, which defines it as

   > "the use of components of biological diversity in a way and at a rate that does not lead to the long-term decline of biological diversity, thereby maintaining its potential to meet the needs and aspirations of present and future generations."

   The CBD definition applies to any form of use, not necessarily only through extraction, but provides clear parameters under which any wildlife harvesting should take place.
3. Much existence value is currently harnessed by way of public contributions to nonprofit organizations that promote wildlife conservation.
4. In economics, a discount rate is the interest rate used to calculate the present value of future cash flows, to determine the current valuation of expected future economic benefits. It represents the opportunity cost of money that could be invested in something that yields a positive return (noting that a current dollar is typically valued more than a future dollar).
5. In recent years there has been an increasing trend in some African countries for government agencies to contract out park management services to nonprofit organizations, including some that have been established to specialize in that function (e.g., the organization African Parks).
6. Also note that government officials have their own private objectives that may differ from public conservation ones (as predicted by public choice theory—see Chapter 1).
7. The incentives for private conservation action are greater than often assumed and are motivated by a range of psychological factors rather than mere financial considerations—see Gooden and Grenyer (2019).
8. See 't Sas-Rolfes et al. (2022) for details of how this has worked in the case of rhino conservation with sustainably managed legal hunting in South Africa and Namibia.
9. Note that the harvester themself may or may not have a significant effect on the price depending on market structure, collusion, etc.
10. Readers with a more advanced knowledge of resource economics will know that the objective function of the harvester can be expressed as a dynamic optimization

problem, which enables seeing clearer links with the rates of extraction and population growth. However, for the sake of simplicity we have chosen to illustrate the concepts here with simple static equations.

11 This will be the case at least until the rate exceeds the sustainable replacement rate and the target population starts to decline. If the decline is sufficiently significant, the harvesting rate will start to decline again too, simply because there is less available stock, and the probability of successful harvest starts to drop. See Copes (1970) for an explanation of this phenomenon, which can lead to a backward-bending long-run supply curve in a fishery, which has also been empirically observed in the abalone industry (Hernández-Casas et al., 2023).

12 Research has also demonstrated that dehorning may have some adverse effects on the behavior of the African black rhino species (Duthé et al., 2023), although no such effects have been observed on the white rhino species.

13 Restrictions on harvesting are also often desirable and put in place to account for the welfare of the affected organisms.

14 Also note that raising the price of the product in this way incurs the risk that any gains may be partly offset by the snob effect discussed in Chapter 4. There is strong evidence of the snob effect existing among consumers—e.g., in Vietnam (see Dang et al., 2020; Dang, 2022).

References

Alchian, A. A., & Demsetz, H. (1973). The property right paradigm. *The Journal of Economic History*, *33*(1), 16–27.

Becker, G. S. (1968). Crime and punishment: An economic approach. *Journal of Political Economy*, *76*(2), 176–177. https://doi.org/10.1086/259394

Biggs, D., Cooney, R., Roe, D., Dublin, H. T., Allan, J. R., Challender, D. W. S., & Skinner, D. (2017). Developing a theory of change for a community-based response to illegal wildlife trade. *Conservation Biology*, *31*(1), 5–12. https://doi.org/10.1111/cobi.12796

Challender, D. W. S., & MacMillan, D. C. (2014). Poaching is more than an enforcement problem. *Conservation Letters*, *7*(5), 484–494. https://doi.org/10.1111/conl.12082

Challender, D. W. S., 't Sas-Rolfes, M., Broad, S., & Milner-Gulland, E. J. (2025). A theory of change to improve conservation outcomes through CITES. *Frontiers in Ecology and Evolution*, *13*. https://doi.org/10.3389/fevo.2025.1425267

Chen, F., & 't Sas-Rolfes, M. (2021). Theoretical analysis of a simple permit system for selling synthetic wildlife goods. *Ecological Economics*, *180*, 106873. https://doi.org/10.1016/j.ecolecon.2020.106873

Clark, C. W. (1973). The economics of overexploitation. *Science*, *181*(4100), 630–634.

Cooney, R., Challender, D. W. S., Broad, S., Roe, D., & Natusch, D. J. D. (2021). Think before you act: Improving the conservation outcomes of CITES listing decisions. *Frontiers in Ecology and Evolution*, *9*. https://doi.org/10.3389/fevo.2021.631556

Copes, P. (1970). The backward-bending supply curve of the fishing industry1. *Scottish Journal of Political Economy*, *17*(1), 69–77. https://doi.org/10.1111/j.1467-9485.1970.tb00487.x

Dang, V. H. N. (2022). When cheap talk is not that cheap – interviewing the super-rich about illegal wildlife consumption. *International Journal of Social Research Methodology, 25*(5), 697–702. https://doi.org/10.1080/13645579.2021.1904117

Dang, V. H. N., & Nielsen, M. R. (2021). Evidence or delusion: A critique of contemporary rhino horn demand reduction strategies. *Human Dimensions of Wildlife, 26*(4), 390–400. https://doi.org/10.1080/10871209.2020.1818896

Dang, V. H. N., Nielsen, M. R., & Jacobsen, J. B. (2020). Reference group influences and campaign exposure effects on rhino horn demand: Qualitative insights from Vietnam. *People and Nature, 2*(4), 923–939. https://doi.org/10.1002/pan3.10121

Duthé, V., Odendaal, K., Van der Westhuizen, R., Church, B., Naylor, S., Boshoff, S., Venter, M., Prinsloo, M., Ngwenya, P., Hanekom, C., Kelly, C. P., Walker, T. W. N., Rasmann, S., & Defossez, E. (2023). Reductions in home-range size and social interactions among dehorned black rhinoceroses (*Diceros bicornis*). *Proceedings of the National Academy of Sciences, 120*(25), e2301727120. https://doi.org/10.1073/pnas.2301727120

Fama, E. F., & Jensen, M. C. (1983). Agency problems and residual claims. *The Journal of Law & Economics, 26*(2), 327–349.

Gooden, J., & Grenyer, R. (2019). The psychological appeal of owning private land for conservation. *Conservation Biology, 33*(2), 339–350. https://doi.org/10.1111/cobi.13215

Gooden, J., & 't Sas-Rolfes, M. (2020). A review of critical perspectives on private land conservation in academic literature. *Ambio, 49*(5), 1019–1034. https://doi.org/10.1007/s13280-019-01258-y

Hernández-Casas, S., Seijo, J. C., Beltrán-Morales, L. F., Hernández-Flores, Á., Arreguín-Sánchez, F., & Ponce-Díaz, G. (2023). Analysis of supply and demand in the international market of major abalone fisheries and aquaculture production. *Marine Policy, 148*, 105405. https://doi.org/10.1016/j.marpol.2022.105405

Hutton, J. M., & Leader-Williams, N. (2003). Sustainable use and incentive-driven conservation: Realigning human and conservation interests. *Oryx, 37*(02). https://doi.org/10.1017/S0030605303000395

Kuiper, T., Haussmann, S., Whitfield, S., Polakow, D., Dreyer, C., Ferreira, S., Hofmeyr, M., Shaw, J., Bird, J., Bourn, M., Boyd, W., Greeff, Z., Hartman, Z., Lester, K., Nowak, I., Olivier, I., Pierce, E., Rowles, C., Snelling, S., … Altwegg, R. (2025). Dehorning reduces rhino poaching. *Science*, 388(6751), 1075–1081. https://doi.org/10.1126/science.ado7490

Laffont, J.-J., & Martimort, D. (2009). *The Theory of Incentives: The Principal-Agent Model*. Princeton University Press.

Ludwig, D., Hilborn, R., & Waters, C. (1993). Uncertainty, resource exploitation, and conservation: Lessons from history. *Science, 260*(5104), 17–36.

McNamara, J., Rowcliffe, M., Cowlishaw, G., Alexander, J. S., Ntiamoa-Baidu, Y., Brenya, A., & Milner-Gulland, E. J. (2016). Characterising wildlife trade market supply-demand dynamics. *Plos One, 11*(9), e0162972. https://doi.org/10.1371/journal.pone.0162972

Meeks, D., Morton, O., & Edwards, D. P. (2024). Wildlife farming: Balancing economic and conservation interests in the face of illegal wildlife trade. *People and Nature*, 6(2), 446–457. https://doi.org/10.1002/pan3.10588

Messer, K. D. (2010). Protecting endangered species: When are shoot-on-sight policies the only viable option to stop poaching? *Ecological Economics*, *69*(12), 2334–2340. https://doi.org/10.1016/j.ecolecon.2010.06.017

Milner-Gulland, E. J., & Rowcliffe, J. M. (2008). *Conservation and Sustainable Use: A Handbook of Techniques*. Oxford University Press.

Mogomotsi, G. E. J., & Madigele, P. K. (2017). Live by the gun, die by the gun: An analysis of Botswana's 'shoot-to-kill' policy as an anti-poaching strategy. *South African Crime Quarterly*, *60*, Article 60. https://doi.org/10.17159/2413-3108/2017/v0n60a1787

Nagin, D. S. (2013). Deterrence in the twenty-first century. *Crime and Justice: A Review of Research*, *42*, 199.

Phelps, J., Carrasco, L. R., & Webb, E. L. (2014). A framework for assessing supply-side wildlife conservation: Wildlife farming and cultivation. *Conservation Biology*, *28*(1), 244–257. https://doi.org/10.1111/cobi.12160

Roe, D., Booker, F., Day, M., Zhou, W., Allebone-Webb, S., Hill, N. A. O., Kumpel, N., Petrokofsky, G., Redford, K., Russell, D., Shepherd, G., Wright, J., & Sunderland, T. C. H. (2015). Are alternative livelihood projects effective at reducing local threats to specified elements of biodiversity and/or improving or maintaining the conservation status of those elements? *Environmental Evidence*, *4*(1), 22. https://doi.org/10.1186/s13750-015-0048-1

Rosenberg, A. A., Fogarty, M. J., Sissenwine, M. P., Beddington, J. R., & Shepherd, J. G. (1993). Achieving sustainable use of renewable resources. *Science*, *262*(5135), 828–829. https://doi.org/10.1126/science.262.5135.828

Soriano-Redondo, A., Alwasiti, H., Kulkarni, R., Correia, R. A., Bryukhova, S., Lita, N. M., Rigor, L. A., Tejerero, D. R., Tenazas, T. M., & Di Minin, E. (2023). Online wildlife trade in species of conservation concern. *Conservation Letters*, *16*(6), e12985. https://doi.org/10.1111/conl.12985

't Sas-Rolfes, M., Challender, D. W. S., & Wainwright, L. (2025). Playing the CITES game: Lessons on global conservation governance from African megafauna. *Environmental Policy and Governance*, *35*(1), 48–63. https://doi.org/10.1002/eet.2123

't Sas-Rolfes, M., & Emslie, R. (2024). African rhino conservation and the interacting influences of property, prices, and policy. *Ecological Economics*, *220*, 108123. https://doi.org/10.1016/j.ecolecon.2024.108123

't Sas-Rolfes, M., Emslie, R., Adcock, K., & Knight, M. (2022). Legal hunting for conservation of highly threatened species: The case of African rhinos. *Conservation Letters*, *15*(3), e12877. https://doi.org/10.1111/conl.12877

't Sas-Rolfes, M., & Gooden, J. (2024). A conflict of visions: Ideas shaping wildlife trade policy toward African megafauna. *People and Nature*, *6*(5), 2029–2045. https://doi.org/10.1002/pan3.10705

't Sas-Rolfes, M., Rademeyer, J., Vigne, L., Emslie, R., Knight, M., Gaymer, J., & Talukdar, B. (2025). The impact of poaching on rhino conservation. In M. Melletti, B. Talukdar, & D. Balfour (Eds.), *Rhinos of the World: Ecology, Conservation and Management* (pp. 367–392). Springer Nature Switzerland. https://doi.org/10.1007/978-3-031-67169-2_15

Tensen, L. (2016). Under what circumstances can wildlife farming benefit species conservation? *Global Ecology and Conservation*, *6*, 286–298. https://doi.org/10.1016/j.gecco.2016.03.007

Thomas-Walters, L., McCallum, J., Montgomery, R., Petros, C., Wan, A. K. Y., & Veríssimo, D. (2023a). Systematic review of conservation interventions to promote voluntary behavior change. *Conservation Biology*, *37*(1), e14000. https://doi.org/10.1111/cobi.14000

Thomas-Walters, L., Morkel, B., Kubo, T., 't Sas-Rolfes, M., Smith, R. J., & Veríssimo, D. (2023b). Understanding the market drivers behind the reduced demand for ivory products in Japan. *Conservation & Society, 21*(1), Article 1.

Torell, E., Crawford, B., Kotowicz, D., Herrera, M. D., & and Tobey, J. (2010). Moderating our expectations on livelihoods in ICM: Experiences from Thailand, Nicaragua, and Tanzania. *Coastal Management, 38*(3), 216–237. https://doi.org/10.1080/08920753.2010.483166

Veríssimo, D. (2013). Influencing human behaviour: An underutilised tool for biodiversity management. *Conservation Evidence, 10*, 29.

Veríssimo, D. (2019). The past, present, and future of using social marketing to conserve biodiversity. *Social Marketing Quarterly, 25*(1), 3–8. https://doi.org/10.1177/1524500419825545

Williamson, O. E. (1983). Organization form, residual claimants, and corporate control. *The Journal of Law and Economics, 26*(2), 351–366. https://doi.org/10.1086/467039

Wright, J. H., Hill, N. A. O., Roe, D., Rowcliffe, J. M., Kümpel, N. F., Day, M., Booker, F., & Milner-Gulland, E. J. (2016). Reframing the concept of alternative livelihoods. *Conservation Biology, 30*(1), 7–13. https://doi.org/10.1111/cobi.12607

9 Toward More Effective Wildlife Trade Regulation

In the previous chapter, we built on the insights of our basic economic model to discuss ways in which the incentives of actors in wildlife trade chains could be influenced for the benefit of conservation. In this chapter, we focus on regulatory reform. Reflecting on the concerns we raised in Chapter 2 about the existing global wildlife trade regime, we consider whether and how the CITES and related approaches could be improved to achieve this goal. We assume that the objectives of the CITES Parties are aligned with the United Nations Sustainable Development Goals, accepting human use of wildlife products within established sustainable limits, as stated in the principles of the widely endorsed Convention on Biological Diversity. Our approach to wildlife trade regulation thus considers its primary goal to be the prevention of unsustainable rates of wild species extraction.

We also assume that society cares about the welfare of individual organisms and that welfare aspects must be considered when establishing wildlife trade policy. The extent to which this ought to happen and be enshrined in law is a subject of much debate, especially as human consciousness of animal sentience continues to develop. In the future, there may be instances in which setting socially acceptable minimum welfare standards might lead to international agreements that preclude any harvesting of some species (e.g., whales),[1] but for now, there is no universally accepted minimum welfare standard that would not allow at least some harvesting of any species—to the extent that we see prohibitions on extraction and trade, they are typically justified as conservation measures. For our approach, we thus consider sustainable extraction rates as the core goal of wildlife trade regulation, subject to minimal welfare standards being met along trade chains.[2]

We divide all species into three loose categories according to the extent of regulatory influence over trade, recognizing that the boundaries between these may be somewhat blurry. The first category applies to species that do not appear in the CITES appendices and for which trade is therefore unregulated (at least, directly) at the international level. The second category applies to species that are listed in either CITES Appendix II or III, which indicates that their trade is at least being somewhat monitored and influenced by regulation.

DOI: 10.4324/9781003286318-9

The third category applies to species listed in CITES Appendix I, for which international trade of wild specimens for commercial purposes is prohibited. Although at least some regulation may apply to all three categories, in the case of the first, it will only be domestic or indirect (e.g., anti-poaching laws, veterinary restrictions, etc.), whereas for the latter two, it is more explicit. Trade may also take place under all three categories; however, for the third (Appendix I-listed species), it is constrained to exceptional conditions such as pre-convention products (e.g., antiques and musical instruments), hunting trophies, and verified captive-bred specimens, all of which are still subject to control through a permitting process before they can cross any international borders.

Unregulated and Novel Species

Our first category concerns species that are not listed in the CITES appendices. This includes the vast majority of species on planet Earth, most of which are also not routinely targeted for harvest and traded by humans. In the case of vertebrate animals, however, an estimated 18% of species may be affected by trade (Scheffers et al., 2019), and many of these remain unregulated. A study of data from 2021 estimated that just over 40% of species noted as "threatened or likely threatened by trade" in IUCN Red List assessments[3] were not listed in the CITES appendices (Challender et al., 2023). Another study observed that some previously overlooked species might draw new attention as potential unregulated substitutes for newly regulated ones through a spill-over effect (Kubo et al., 2025). Furthermore, as humanity's global footprint increases, new species and uses are discovered, drawing potential trade interest. Finally, various rapid technological advances increase the risks of rapidly expanding the identification and exploitation of newly popularized species (Esmail et al., 2020).

If trade in this category of species is observed, it should be rapidly assessed for potential links to unsustainable levels of extraction. For many species, this may be challenging, as population levels are not always easily monitored. However, if trade is legal, it should be relatively easy to obtain and monitor price data, which can provide clues of an emerging threat—rapidly rising prices are a red flag, likely indicating increasing relative scarcity. Without price data for reference, data on quantities of wildlife products alone do not necessarily provide a reliable indicator of an emerging problem unless the quantities are high in absolute terms compared with the estimated species population size and growth rate. Sustainability assessments should consider both ecological and economic factors, separating suspected trade-related threat drivers from others and focusing on wild population growth rates in relation to targeted extraction.

Ideally, all wild species should be physically protected from uncontrolled and excessive extraction at their source. Wildlife trade regulation is generally a poor, second-best substitute for effective *in situ* conservation measures.

As discussed in the previous chapter, before resorting to trade measures, conservationists should explore the potential for institutional reforms that create stronger incentives for local people living with wildlife to become effective custodians (e.g., stronger property rights regimes). The best hope for any species is to increase the probability that potential poachers are intercepted before they even reach the organisms they are seeking to harvest, such that they are deterred from attempting to do so in the first place. In some instances, this might mean creating systems that enable local people to benefit from some (legal) sustainable harvest and use (and possibly trade) of wildlife products. However, the potential for this should be carefully evaluated for each specific case.

If an emerging conservation issue is suspected and there is limited scope for improving the situation at source, consumer engagement should ideally precede regulation as the next step. Identifying and communicating with consumers can provide critical information to decide on the most appropriate strategy. The extent to which consumers are willing to accept substitutes or commit to avoiding unsustainable sources will determine whether a facilitated legal market (potentially including introduced substitutes) or tighter restrictions are more appropriate. Once a species has been flagged for more punitive regulation, this has two potentially adverse effects: (1) consumers may start to conceal their behavior and be less forthcoming with important information and (2) consumers may be motivated to start stockpiling the product in the expectation of future scarcity, thereby causing accelerated wild harvesting of the species (see Rivalan et al., 2007). As critical stakeholders, consumers should be co-opted to help tackle the problem before they are made part of the problem.

If, after consulting with both relevant local people and consumers, regulation is considered appropriate, it should be planned in consultation with all identified actors along the trade chain to determine whether an appropriately monitored and traceable trading system can be developed. If trading stakeholders collaborate with conservation management authorities to develop a mutually acceptable system, this has the greatest chance of averting a conservation problem. Any system should be designed with the market structure in mind, considering the extent to which monopoly power might exist along the trade chain and whether that should be mitigated.[4] The agreed system could include a CITES listing to support a certification mechanism. Alternatively, if trading stakeholders are unwilling to cooperate, stricter regulation could be imposed, mindful of whether the capacity to enforce it exists in all relevant jurisdictions along the trade chain.

Trade-Regulated Species

Our second category concerns species listed on CITES Appendices II and III, for which international trade is permitted but at least somewhat regulated. Unlike the previous category, there is at least some existing awareness that the species may be threatened by extraction for trade. For these species, the question to consider is whether regulations could be improved to ensure their effectiveness. Ideally, this requires evaluation on a case-by-case basis. Examining a case would start with a sustainability review, such as the existing "Review of Significant Trade" process that takes place within CITES.[5] Evaluations should follow similar principles as those of unregulated species, i.e., they should examine trends in both biological and economic data (e.g., market prices), noting the limitations of existing data sources, including the official CITES Trade Database and LEMIS (see Challender et al., 2022).

As with unlisted species, addressing *in situ* conditions often provides the key to resolving sustainability concerns. Any attempts to modify trade regulation should consider how these may affect the institutional arrangements and related net benefit flows that would incentivize local people to conserve rather than overexploit. They should also consider the likely responses of consumers, their willingness to change their purchasing habits, and the likely responses of existing intermediaries. Such responses may impact product prices and ultimately influence the incentives for extraction (and possible aggravating factors such as speculative stockpiling).

Modifications to trade regulations may either move toward easing or toward further restriction (e.g., by changing listing status, setting quotas, or amending more specific domestic permitting requirements). In practice, easing through downlisting is seldom considered as it requires a decisive improvement in the conservation status of the species. However, if appropriate supporting institutions are established, easing restrictions may be desirable to open more profitable avenues of sustainable conservation financing.

Assessments of existing regulations should also consider how they influence competition from other sources (e.g., wildlife farming) and their longer-term effects on industry structure and strategic behavior of market actors. Unsustainable harvesting might be mitigated by regulatory change that eases the trade of substitutes relative to wild-sourced products; however, laundering potential and incentives must be considered, as well as animal welfare concerns in the case of farmed animal product substitutes. These specific concerns will likely differ between species, along with consumer responses to such changes (and therefore prices of the various products). Conversely, if increased restrictions are considered, regulators should ensure that they are proportionate[6] and do not cause perverse overreactions that undermine their intent—the risks of excessive regulation[7] should not be underestimated.

In all cases, amendments to regulatory approaches should attempt to improve monitoring of the status of wild populations (growth and extraction rates) and market responses (e.g., quantities traded and prices) along trade chains. However, when species are uplisted to CITES Appendix I, monitoring of market responses is compromised as any persisting international commercial trading activity becomes clandestine. As a further potential disadvantage, uplisting may signal increasing future scarcity, and stimulate increased speculative demand and intensified poaching efforts in the lead-up to the new regulations being fully implemented and enforced. These factors—and the longer-term prospects of species recovery to the point of potential future downlisting—should be carefully considered. Appendix I listings should be understood as a sign of species conservation failure, not viewed as a victory (Challender & MacMillan, 2019).

CITES Appendix I-Listed Species

Species listed on CITES Appendix I are effectively subjected to prohibition in international trade for commercial purposes (and thus of any substantial volume). Such prohibition typically aims to protect the subject species by impeding access to significant foreign consumer markets. Appendix I listings raise the transaction costs of trading between countries, which may or may not succeed in reducing the rate of harvesting for trade to sustainable levels, depending on the supply and demand functions (and related incentive structures) within the prohibition regime. For many wildlife products, consumer demand may persist despite the listing, in which case illegal trade will continue as the only mechanism to supply the market. Criminal intermediaries thus gain market power, as they no longer face competition from legal traders. The extent to which they gain such power—and the consequences thereof—will again vary with species and circumstance.

In cases of wildlife products for which foreign consumer demand is somewhat limited and elastic (i.e., consumers can readily switch to substitute products), trade bans can raise the transaction costs to levels that stifle market profitability sufficiently to reduce illegal extraction rates to sustainable levels. However, in cases for which demand and supply are less elastic (similar to markets for other products, such as some illegal drugs[8]), poaching and illegal trade may persist at threatening levels, as is the case with rhinos and elephants in many parts of their range (Di Minin et al., 2022). In such instances, there are two distinct policy options. The first is to pursue the approach previously adopted for crocodilians and the vicuña and develop a strategic plan to reintroduce appropriately structured and managed legal trade to benefit the species concerned, consistent with a socioecological harm reduction approach.[9] The second is to maintain the Appendix I listing and attempt to further intensify regulatory, punitive, and enforcement measures.

Since the inception of CITES, and especially in more recent years, the second approach has been increasingly adopted for numerous species, notably high-profile examples such as rhinos, elephants, big cats, and pangolins. Following the early twentieth-century surges in poaching of many of these species (attributed to rising consumer demand in East Asian markets), there has been a growing trend to frame wildlife trafficking as an international security issue, drawing in a broader range of international organizations[10] with the aim of combatting and disrupting transnational organized crime. This approach has included various forms of international cooperation, such as employing anti-money laundering measures. However, despite some high-profile arrests and seizures, it remains unclear whether this approach is delivering significant measurable conservation success or that it can deliver in the future, given that similar but longer-standing attempts to tackle the ongoing international trafficking of drugs[11], arms, and humans have yielded relatively disappointing results. As mentioned in Chapter 1, various scholars have expressed concerns over the consequent militarization of conservation, and there are further practical and philosophical arguments against "over-criminalization" as a matter of principle (Husak, 2008).

In a similar vein, efforts to fully implement trade bans may also be misguided from a conservation perspective, notably when they target existing stocks of storable products from sources other than recent poaching[12]—if such products find their way into markets, they may suppress prices and thereby reduce incentives for fresh poaching (a conservation benefit). As discussed in Chapter 2, if some consumer demand persists, rigorously maintaining long-term trade bans creates a risk that the species concerned may eventually become trapped in a systemic "poaching pit" whereby all attempts at intensified prohibition drive up product prices to a greater extent than costs of illegal supply, which in turn results in intensified poaching efforts, supported and enabled by illegal trade innovations and increased corruption, thus preventing species populations from regaining abundance in large open systems (in which both access control and early detection opportunities are constrained). In this scenario, high-value species may increasingly become confined to areas of more intensive management[13] with a greater risk of eventual extinction in the wild.

Radical Regulatory Reform

Finally, we briefly consider options for more radical reform of the current global regulatory regime for wildlife trade. As discussed in Chapter 2, numerous critics have expressed concern for decades that CITES is not sufficiently effective in achieving its stated conservation objectives. This may be partly due to polarizing ideological tensions within the CITES community, entrenched by the treaty's institutional structure, which is based on the differential appendix listing system and decisions ultimately determined through voting ('t Sas-Rolfes & Gooden, 2024; 't Sas-Rolfes et al., 2025). This enables the CITES system to be gamed by special interest groups, potentially leading to suboptimal outcomes in both conservation and socioeconomic terms. The CITES regulatory logic also implies a theory of change[14] with questionable assumptions (Challender et al., 2025), especially concerning the incentives and ability of governments to implement and enforce all the supporting regulations. Supporters of a more economic approach to regulation thus advocate investigating reforms to CITES to better align actor incentives along trade chains.[15]

Conversely, supporters of the more legalistic approach to CITES increasingly advocate a more radical alternative called "reverse listing" (Ditkof,

1982; Marshall et al., 2025) or "whitelisting" (Couzens, 2013). Proponents of this approach argue that the existing system of listing species to be protected through trade regulation is reactive (rather than proactive) and subject to detrimental time lag effects. They propose shifting the burden of proof (of sustainability) onto those wishing to trade by effectively banning trade in all species other than those listed as positively approved for trade (possibly subject to a quota). Trade in any species would thus only be permitted once the CITES Parties have evaluated and approved a trading proposal.

In theory, such a system might have certain advantages, such as potentially simplifying cross-border trade enforcement and eliminating the problem of delayed listings, especially of novel species (Macdonald et al., 2021). However, in practice, changing the current system presents both serious implementation challenges and risks. It is unclear (and seems unlikely) that the CITES Parties would readily agree to a sudden radical shift in approach, given the extent to which their existing policies and procedures would be disrupted.[16] For countless traded wild species, this policy change would represent a sudden shift from a legal trading regime to a prohibition regime, which could create many of the problems discussed above (e.g., increasing the market power of certain legal and illegal actors, prompting speculative stockpiling, and curtailing legal benefit flows to local conservation initiatives and communities). There is a substantial risk that such a large-scale change would overwhelm the capacity of official conservation and other government agencies while providing an advantage to more nimble criminal entrepreneurs.

CITES and the global wildlife trading regime have grown and evolved substantially during the five decades since the treaty was ratified. However, its evolution has been largely incremental, with just a few observable inflection points. There can be no doubt that the treaty has its shortcomings—we believe that this is in large part due to an overreliance on a simplistic regulatory logic that fails to adequately consider economic factors and second-order effects. A move to "reverse listing" would overcome few of the existing shortcomings and potentially aggravate some of them. However, there appears to be scope for some reasonably fundamental institutional reform within the regime, and this is a topic worthy of further research.

Notes

1 For useful and insightful discussions on the emergence of ethical concerns relating to animal welfare and their potential role in future international wildlife law, see Scholtz (2017a; 2017b).

2 We avoid prescribing what welfare standards might be appropriate, as our focus in this book is on conservation. However, we recognize and support the validity of setting and maintaining standards of humane treatment of nonhuman organisms, also aware of instances in which the simultaneous pursuit of welfare and conservation objectives involves facing challenging trade-offs.

3 The IUCN Red List of Threatened Species is the world's most comprehensive information source on the global extinction risk status of wild species. Assessments

are conducted by scientists through a global, objective process, based on a well-established, quantitative protocol.

4 Monopoly power of profit-seeking entities at the intermediary level requires particular attention, as explained in Chapters 4 and 7.
5 See https://cites.org/eng/imp/sigtradereview for further information.
6 As an example of disproportionate restrictions, see the proposed UK Hunting Trophies (Important Prohibition Bill (Challender et al., 2024).
7 For further detailed discussion on appropriate regulatory design, see Baldwin et al. (2013) and Drahos (2017), especially the contribution by Gunningham and Sinclair (2017) on "smart regulation."
8 See Becker et al. (2006).
9 Successfully legalizing trade of a previously banned wildlife product while ensuring no detriment to the species is challenging and would require careful planning to mitigate potential strategic responses from existing illegal actors—in most cases a mere CITES downlisting may be too risky without specific supporting measures. The most effective approach will vary significantly with species, geography, product market characteristics, and other factors, and should include careful consideration of the lessons from our economic model (e.g., role of market structure, substitutes, stockpiling incentives, etc.). However, presenting any sort of standard blueprint for legalization is beyond the scope of this book.
10 Examples include the United States Bureau of International Narcotics and Law Enforcement, the Royal Foundation's United for Wildlife initiative, the Global Initiative Against Transnational Organized Crime, and the International Consortium on Combatting Wildlife Crime, a partnership between the CITES Secretariat, INTERPOL, United Nations Office on Drugs and Crime, the World Bank, and the World Customs Organization.
11 In December 2024, the United Nations High Commissioner for Human Rights, Volker Türk, denounced the decades-long global "War on Drugs" as a complete and utter failure, noting that this approach had destroyed countless lives and damaged entire communities. He called for a radical shift in approach to prioritize health, dignity, and inclusion.
12 Examples include attempts to prevent old stocks of elephant ivory in Japan from being traded to China and accumulated stocks of big cat bones from lion and tiger breeding facilities from "leaking" into markets.
13 This has happened with some rare antelope and plant species (e.g., cycads) and appears to be the trend for rhino species too.
14 A theory of change is a framework that outlines how and why an approach is expected to lead to specific outcomes and impacts, essentially mapping out the causal pathway from activities to desired results.
15 There are various ways in which this may be done, some requiring more radical amendments to treaty than others. We regard making detailed recommendations as beyond the scope of this book.
16 Serious consideration of such reform ought to be preceded by a comprehensive survey of the Parties with the largest trade volumes on both the supply and demand side (i.e., both producer and consumer countries) to enable them to assess the potential impact on their wildlife populations and markets.

References

Baldwin, R., Cave, M., & Lodge, M. (2013). *Understanding Regulation: Theory, Strategy, and Practice* (2nd ed.). Oxford University Press.

Becker, G. S., Murphy, K. M., & Grossman, M. (2006). The market for illegal goods: The case of drugs. *Journal of Political Economy, 114*(1), 38–60. https://doi.org/10.1086/498918

Challender, D. W. S., Brockington, D., Hinsley, A., Hoffmann, M., Kolby, J. E., Massé, F., Natusch, D. J. D., Oldfield, T. E. E., Outhwaite, W., 't Sas-Rolfes, M., & Milner-Gulland, E. J. (2022). Mischaracterizing wildlife trade and its impacts may mislead policy processes. *Conservation Letters, 15*(1), e12832. https://doi.org/10.1111/conl.12832

Challender, D. W. S., Cremona, P. J., Malsch, K., Robinson, J. E., Pavitt, A. T., Scott, J., Hoffmann, R., Joolia, A., Oldfield, T. E. E., Jenkins, R. K. B., Conde, D. A., Hilton-Taylor, C., & Hoffmann, M. (2023). Identifying species likely threatened by international trade on the IUCN Red List can inform CITES trade measures. *Nature Ecology & Evolution, 7*(8), 1211–1220. https://doi.org/10.1038/s41559-023-02115-8

Challender, D. W. S., & MacMillan, D. C. (2019). Investigating the influence of non-state actors on amendments to the CITES appendices. *Journal of International Wildlife Law & Policy, 22*(2), 90–114. https://doi.org/10.1080/13880292.2019.1638549

Challender, D. W. S., 't Sas-Rolfes, M., Dickman, A., Hare, D., Hart, A. G., Hoffmann, M., Mallon, D., Mandisodza-Chikerema, R. L., & Roe, D. (2024). Evaluating key evidence and formulating regulatory alternatives regarding the UK's Hunting Trophies (Import Prohibition) Bill. *Conservation Science and Practice, 6*(10), e13220. https://doi.org/10.1111/csp2.13220

Couzens, E. (2013). CITES at forty: Never too late to make lifestyle changes. *RECIEL, 22*(3), 311–323.

Di Minin, E., 't Sas-Rolfes, M., Selier, J., Louis, M., & Bradshaw, C. J. A. (2022). Dismantling the poachernomics of the illegal wildlife trade. *Biological Conservation, 265*, 109418. https://doi.org/10.1016/j.biocon.2021.109418

Ditkof, M. (1982). International trade in endangered species under C.I.T.E.S.: Direct listing vs. reverse listing. *Cornell International Law Journal, 15*(1), 107–120.

Drahos, P. (2017). *Regulatory Theory: Foundations and applications*. ANU Press.

Esmail, N., Wintle, B. C., 't Sas-Rolfes, M., Athanas, A., Beale, C. M., Bending, Z., Dai, R., Fabinyi, M., Gluszek, S., Haenlein, C., Harrington, L. A., Hinsley, A., Kariuki, K., Lam, J., Markus, M., Paudel, K., Shukhova, S., Sutherland, W. J., Verissimo, D., … Milner-Gulland, E. J. (2020). Emerging illegal wildlife trade issues: A global horizon scan. *Conservation Letters, 13*(4), e12715. https://doi.org/10.1111/conl.12715

Gunningham, N., & Sinclair, D. (2017). Smart regulation. In P. Drahos (Ed.), *Regulatory Theory* (pp. 133–148). ANU Press; JSTOR. www.jstor.org/stable/j.ctt1q1crtm.16

Hiller, C., & 't Sas-Rolfes, M. (2025). Systematic review of the impact of restrictive wildlife trade measures on conservation of iconic species in southern Africa. *Conservation Biology, 39*(1), e14262. https://doi.org/10.1111/cobi.14262

Husak, D. (2008). *Overcriminalization: The Limits of the Criminal Law* (1st ed.). Oxford University Press.

Kubo, T., Mieno, T., Uryu, S., Terada, S., & Veríssimo, D. (2025). Banning wildlife trade can boost the unregulated trade of threatened species. *Conservation Letters, 18*(1), e13077. https://doi.org/10.1111/conl.13077

Macdonald, D. W., Harrington, L. A., Moorhouse, T. P., & D'Cruze, N. (2021). Trading animal lives: Ten tricky issues on the road to protecting commodified wild animals. *BioScience, 71*(8)), 846–860. https://doi.org/10.1093/biosci/biab035

Marshall, B. M., Alamshah, A. L., Cardoso, P., Cassey, P., Chekunov, S., Eskew, E. A., Fukushima, C. S., García-Díaz, P., Gore, M. L., Lockwood, J. L., Rhyne, A. L., Sinclair, J. S., Strine, C. T., Stringham, O. C., Tlusty, M. F., Valdez, J. W., Watters, F., & Hughes, A. C. (2025). The magnitude of legal wildlife trade and implications for species survival. *Proceedings of the National Academy of Sciences, 122*(2), e2410774121. https://doi.org/10.1073/pnas.2410774121

Rivalan, P., Delmas, V., Angulo, E., Bull, L. S., Hall, R. J., Courchamp, F., Rosser, A. M., & Leader-Williams, N. (2007). Can bans stimulate wildlife trade? *Nature, 447*(7144), Article 7144. https://doi.org/10.1038/447529a

Scheffers, B. R., Oliveira, B. F., Lamb, I., & Edwards, D. P. (2019). Global wildlife trade across the tree of life. *Science, 366*(6461), 71–76. https://doi.org/10.1126/science.aav5327

Scholtz, W. (2017a). Injecting compassion into international wildlife law: From conservation to protection? *Transnational Environmental Law, 6*(3), 463–483. https://doi.org/10.1017/S2047102517000103

Scholtz, W. (2017b). Killing them softly? Animal welfare and the inhumanity of whale killing. *Journal of International Wildlife Law & Policy, 20*(1), 18–37. https://doi.org/10.1080/13880292.2017.1309865

't Sas-Rolfes, M., Challender, D. W. S., & Wainwright, L. (2025). Playing the CITES game: Lessons on global conservation governance from African megafauna. *Environmental Policy and Governance, 35*(1), 48–63. https://doi.org/10.1002/eet.2123

10 Conclusions

In this book, we have discussed the importance of the wildlife trade in relation to conservation and highlighted ways to think about it that extend beyond conventional thinking on the topic. We started by noting that the definitions of wildlife and trade are not always clear and that there are consequently some misgivings about the nature and role of wildlife trade. There are both benefits and issues of social concern relating to the trade in wildlife products. We then explained how wildlife conservation issues have a strong economic dimension and that we can draw many valuable lessons from the history of economic thought to ultimately help us conserve wildlife more effectively. Using the example of rhino trade regulation, we also explained how past economic ignorance resulted in poor trade policy decisions, which is why we are still dealing with a serious rhino poaching problem after 50 years of CITES-linked attempts to stop this.

We continued to explain that conservation is not a straightforward issue—it is about more than simply preventing species from going extinct. Because there are many different ethical and practical approaches to conservation, especially in the context of the Anthropocene, there are also conflicting attitudes toward market approaches, including wildlife trade. We discussed the importance of institutions in shaping human behavior and how the notion of property plays a vital role in influencing wildlife conservation and trade outcomes in relation to markets and government policies. After also highlighting the pivotal role of entrepreneurs, we discussed theoretical and practical aspects of wildlife trade regulation, highlighting the differences between a conventional legalistic approach that assumes that governments will simply implement laws successfully and an economic approach that takes a more critical view, keeping the incentives of market participants in mind.

We then examined existing wildlife markets, legal and illegal, their structures, dynamic nature, and participants. We considered a wide range of species in trade and their products before explaining how initial simple transactions between a harvester and a consumer could eventually evolve into complex and sophisticated markets. We described how trade chains could develop and how market structures and processes might emerge in

DOI: 10.4324/9781003286318-10

various ways, drawing from examples from four product market categories. This discussion led us to consider the types of active market participants in the trade chain—the actors—and the incentives they face, given a common desire to benefit from trade. All actors weigh up costs and benefits before deciding to trade at a specific price, and this provides the basis for our analytical economic framework.

We proceeded to present our basic economic model and demonstrated how it could be extended to examine the potential impact of various circumstances and specific policy interventions. We then considered how our insights could be applied toward influencing the incentives of various actors within wildlife trade systems. We further drew on our accumulated insight to consider whether and how existing approaches to wildlife trade regulation may be improved for the benefit of conservation.

An analysis of wildlife trade ultimately comes down to issues of supply and demand: for example, how are the wildlife goods sourced?; how much of them are available?; how much does it cost suppliers to put them out on the market?; how much do buyers value these goods; are they willing to buy close substitutes?; if so, how much would they pay for them?; and other related questions. These are all within the purview of economics. Understanding wildlife goods markets and how to regulate them for the betterment of society is therefore as much an economics problem as it is a conservation problem.

In this book, we have thus presented a rigorous analytical framework built on economic principles to examine markets for wildlife goods. We used our models to look at the potential effects of, for instance, various conservation policies, introducing a market for substitute goods, and the existence of speculators who stockpile wildlife goods for resale purposes. A major theme that emerges from our analysis is that market structure—i.e., whether the markets in question are competitive or dominated by a few major players—matters a great deal in terms of how the markets operate. Another vital theme is that how much information consumers possess regarding what type of wildlife goods—be they wild-sourced, synthetic substitutes, or derived from farmed animals—they are getting is critical in determining the market equilibrium—and even how many equilibria there are in the market. These lessons, naturally, have significant implications for the conservation of wildlife species that are threatened or endangered by commercial trades involving them.

The most crucial lesson from the book, we believe, is the importance of having a rigorous modeling framework based on solid economic foundations to examine issues involving wildlife goods markets. This framework allows us to better understand the incentives facing the various players in these markets, shows how different market forces interact and play out to determine the market equilibrium, and gives us a tool with which to perform hypothetical policy experiments in these markets.

Several caveats regarding our results and analyses are in order. Even though we have demonstrated in certain contexts how important market structure is in determining the market outcome, we have barely scratched the surface of this topic. For simplicity, we have only made comparisons between perfectly competitive markets on the one hand and markets controlled by a monopoly on the other hand. Future in-depth expositions on this issue would need to consider different forms of market structure and examine in greater detail the connection between how markets are organized and how much (and what kinds of) wildlife goods are traded.

Thorough analyses of wildlife trade, especially in the context of conservation, must be dynamic in nature. For example, how much wildlife is harvested today to produce goods for sale will in general affect how much wildlife can be harvested in the future (and how much it costs to harvest them). In addition, how much wildlife goods are sold today could alter how much value consumers place on those products later. These inter-temporal trade-offs affect how market participants behave, and they would have to be accounted for by policymakers when considering what policies to implement and what institutions to design for regulating markets for wildlife goods.

For ease of exposition and simplicity, we have largely omitted the dynamic aspects of wildlife goods markets in our analyses except in Chapter 7, which gives an exposition of speculators and stockpiling, and where the dynamic components are indispensable. Even there, we have presented a highly simplified model that incorporates only two time periods. We note here that many questions concerning stockpiling remain unanswered—such as how introducing a market for close substitute goods would impact speculators—which means there is fertile ground here for conservation economists to work with in the future.

As mentioned briefly earlier, consumers' preferences or liking for wildlife goods may shift over time or change under certain conditions. For instance, when a substitute good is made available, people's willingness to pay for a wild good may be positively or negatively affected. Although we touched on changing preferences in Chapter 4 when discussing the snob effect, we assumed throughout the rest of the book that consumer preferences remain stable (in Chapter 7 in particular). We emphasize, however, that the economic framework presented here can easily accommodate changing valuations and preferences, i.e., the models do not need stable preferences to work. Ultimately, under what conditions do consumers' preferences change—and how much—is an empirical question, and future research can investigate how estimated changes in preference can affect the results and lessons derived here. In this vein, the analyses presented here that assume stable preferences can be considered benchmark cases against which other models and variations can be compared.

We have used the models here to examine the potential effects of several conservation policies, such as demand-side promotion of a substitute good, implementing more effective anti-poaching measures, and introducing a permit system for selling a substitute good. We note that it is not our intention to study all existing or proposed conservation policies—for that would be beyond the scope of this book—and our choice of what policies to consider in the book does not necessarily reflect what we believe to be the best or most effective policies to use for conservation purposes. Our choices come down to the fact that these are frequently discussed policy options—and not some far-fetched esoteric proposals. In addition, they serve as good examples for us to use to illustrate how the economic models work. A primary objective of our book is to show how to construct economic models of wildlife goods markets and how to use them in analysis—it is emphatically not a comprehensive or exhaustive analysis of conservation policies. Proposals or issues not considered here could be accommodated and studied in the same models as the ones presented here or in variations or extensions of these models. In other words, the economic analysis of wildlife goods markets is not restricted to the topics and policies studied here.

Issues of social welfare—especially those that include the welfare of animals or plants, which could be at the heart of discussions surrounding trade in wildlife goods—are inherently complex since it is difficult to quantify the well-being of living things. When the calculus of conservation pits one wildlife species against another (e.g., should we save spotted owls at the expense of barred owls?), the economic framework we have put forth does not help us in resolving the issue, since it is silent regarding how we assign values to different species. While numerous treatises and scholarly works attempt to impute values to various aspects of the biological and ecological worlds to help arrive at some "optimal" policy prescription (as discussed in Chapter 2), that is not our intention here. Where the economic way of thinking is helpful, however, is in identifying inefficiencies—situations in which one or more parties can be made better off without hurting any other parties—and in predicting market outcomes. It would be a fruitless exercise to talk about society's welfare from a particular set of policies without being able to forecast how markets for wildlife goods would react to those policies. And it would be a win-win situation for everyone involved if the economic models could be used to find policy solutions that could eliminate any inefficiencies that exist and make some participants (including wildlife) better off without hurting others.

It would be remiss of us if we did not discuss how "realistic" the economic models are, i.e., how closely they represent some specific wildlife goods markets. There are two ways to make use of these formal models. One way is to construct models that—balancing realism and tractability and using empirical data where available and appropriate—capture some particular markets as closely as possible in terms of certain features. The main benefit of this 'literal' approach to modeling is the ability to derive specific quantitative predictions regarding the equilibria of these markets and how they are impacted by changes in model parameter values.

The other approach, which we could call the 'figurative' approach of modeling, sacrifices realism in favor of greater generality. In this approach, the modeler identifies certain key qualities of the markets or situations in question and constructs the simplest model that exhibits these key properties.

The goal is qualitative rather than quantitative: for example, to identify general conditions under which some outcomes occur rather than to ask how much a variable would change in response to a certain amount of change in another variable. The exposition in Chapter 6 on adverse selection is an example of this figurative approach. We did not attempt to replicate features of any specific wildlife goods market in that discussion. Yet, by taking as our starting point certain key features of markets with asymmetric information—such as the inability of consumers to distinguish between a good and its close substitute—we can derive general results that apply to any market with these features.

The two approaches of using models are not competing—they are used for different purposes, and each has its own strengths and weaknesses. In our book, we have taken a figurative approach, for we are mainly interested in showing how the economic framework is used and what kinds of outcomes could occur in wildlife goods markets in general without getting into the nitty-gritty of the literal approach. For research and policy questions that are quantitatively oriented, the literal approach would be required, and an in-depth discussion or exposition of this is outside the scope of this current book.

To make our exposition and analyses accessible to our target audience—conservation scholars, stakeholders, and policymakers—who may not have much of a background in economics beyond an introductory course—we have attempted to make the models as simple as we could, and we have used a graphical approach without heavy mathematics whenever possible. It is difficult, however, to escape the fact that an adequate understanding of many issues surrounding wildlife trade—such as the importance of market structure, the effects of incomplete information in the market, and the dynamics of speculative stockpiling—requires much more knowledge of economics than what is typically taught in an Economics 101 class. We therefore sincerely hope that this book will stimulate interest in the study of economics, encourage more scholars to engage in conservation issues with the aid of an economic framework, and help policymakers and various stakeholders take actions with a more holistic and nuanced understanding of how wildlife goods markets operate.

Index

Note: Endnotes are indicated by the page number followed by 'n' and the endnote number e.g., 20n1 refers to endnote 1 on page 20.

For Product Safety Concerns and Information please contact our EU representative GPSR@taylorandfrancis.com
Taylor & Francis Verlag GmbH, Kaufingerstraße 24, 80331 München, Germany

www.ingramcontent.com/pod-product-compliance
Ingram Content Group UK Ltd.
Pitfield, Milton Keynes, MK11 3LW, UK
UKHW020002200926
475702UK00015B/2424
* 9 7 8 1 0 3 2 2 6 0 5 3 2 *